SMART PRACTICES

Introductory Ordinary Differential Equations

Including Ten Fully Solved Practice Examinations

Peter Schiavone
University of Alberta

Canadian Cataloguing in Publication Data

Schiavone, Peter, 1961-
Introductory ordinary differential equations

(Smart practices)
ISBN 0-13-907338-8

1. Differential equations. I. Title. II. Series.

QA372.S34 1998 515'.352 C97-932281-2

To Linda and Francesca

Prentice-Hall, Inc., Upper Saddle River, New Jersey
Prentice-Hall International (UK) Limited, London
Prentice-Hall of Australia, Pty. Limited, Sydney
Prentice-Hall Hispanoamericana, S.A., Mexico City
Prentice-Hall of India Private Limited, New Delhi
Prentice-Hall of Japan, Inc., Tokyo
Simon & Schuster Southeast Asia Private Limited, Singapore
Editora Prentice-Hall do Brasil, Ltda., Rio de Janeiro

ISBN 0-13-907338-8

Publisher: Patrick Ferrier
Acquisitions Editor: Sarah Kimball
Editorial Assistant: Melanie Meharchand
Director of Marketing: Tracey Hawken
Production Editor: Andrew Winton
Production Coordinator: Jane Schell
Cover Design: David Cheung

1 2 3 4 5 WC 02 01 00 99 98

Printed and bound in Canada.

CONTENTS

PREFACE

This publication is intended for all students enrolled in any program which requires an introductory course in *ordinary differential equations*. It is written specifically to appeal to students and to motivate them to succeed in course examinations.

Examinations invariably make up the largest component of a student's final grade in a typical mathematics course. Consequently, most students are focused on doing well *in the examinations* rather than on an appreciation of the course material. In this book, the emphasis is on *maximizing performance* with particular reference to *course examinations*.

Examinations in introductory *ordinary differential equations* courses are largely predictable - mainly because of the limited range and repetitive nature of the topics covered. As a result, the same *types* of questions appear over and over again. Once the most relevant and significant areas of expertise have been identified, the task reduces to working through as many *relevant* problems as possible: problems similar to those likely to be asked in the actual examinations, preferably practice-examination questions. In doing so, students become *fluent* in the relevant techniques and begin to see repetition and *patterns* in the solutions. Eventually the solution process becomes so systematic that the actual examination becomes an anticlimax. This is how the best students succeed in mathematics courses.

There are two key ingredients in this procedure:

(i) A good supply of practice-examination questions.

(ii) Full, easy-to-read *solutions* to every problem.

The latter make the difference! It is not difficult to obtain old examinations but fully-worked, comprehensive solutions are almost never available. Even when they are available, they are usually not well-presented and often confusing.

The book is divided into two main parts. Part 1 is concerned with a step-by-step review of the *essentials* from a basic course in ordinary differential equations. Here, I review the main techniques used to solve first and higher order differential equations and illustrate each technique with a fully worked example. The material in Part 1 is *specific* and *targeted*, the emphasis being on providing a reference for the material used in deriving the solutions to the examinations in Part 2. Part 1 will serve also as a valuable reference for specific formulae and techniques *during* and *after* a course in ordinary differential equations.

Part 2 of the book begins with five practice midterm examinations from a typical course in introductory ordinary differential equations. These are followed by five practice final examinations and, finally, detailed solutions to all ten examinations. The solutions are written more to provide teaching assistance than to furnish a set of answers. In each solution, I include the relevant theory and formulae and provide a reference to the appropriate section in Part 1 of the text. The reason for this comes from my own experience as an instructor: students learn most when applying the theory to examples, particularly when they have either someone to ask or a full set of *teaching solutions*. I recommended the solutions be studied carefully, in particular, the steps and the reasoning behind each solution. Practice in this context will help develop a clear and logical approach to each *type* or *class* of problem.

The examinations are rated on a scale of 1 (easy) to 5 (difficult) and include information such as mark distribution and allocated time. Only examinations of difficulty levels 3 to 5 are addressed in this text.

I have also included an Appendix of the most important supplementary techniques from Calculus (for example, the techniques of integration). This makes the book essentially self-contained.

Finally, I should mention that this particular publication is intended to supplement or complement, rather than replace, the regular offerings of an introductory course in Ordinary Differential Equations. My intention is not to "re-teach" the material, but to provide valuable supplementary information (in the form of solved practice examinations) aimed at enhancing performance in a course which is common to students of engineering, business, and science-based subjects, alike. For this reason, the theory and techniques in Part 1 are presented without proofs and with little explanation (other than illustrative examples). I leave that to the many excellent textbooks on the subject. Rather, Part 1 should be used mainly as a reference to supplement Part 2.

I recommend this book be used particularly when *preparing* for course examinations. Apart from *sharpening* existing skills and *fine-tuning* exam preparation, the material will help judge levels of comprehension and preparedness.

(Note that the symbol § is occasionally used to identify a reference to a chapter section.)

Peter Schiavone, Ph.D.

INTRODUCTION

Part 1 covers the full range of topics offered in a regular introductory course in ordinary differential equations. In order, these are:

1. **Terminology, notation and basic concepts associated with the theory of differential equations.**

2. **Solution Techniques for first order ordinary differential equations:**

   - Separable equations
   - Homogeneous equations
   - Exact equations
   - Solution by integrating factors
   - Linear equations
   - Bernoulli equations
   - General strategy for solving first order equations
   - Applications - orthogonal trajectories

3. **Solution Techniques for Higher Order Differential Equations**

   - Theory of linear ordinary differential equations
   - Linear homogeneous differential equations with constant coefficients
   - Linear inhomogeneous differential equations with constant coefficients :
     The method of Undetermined Coefficients
   - Reduction of order
   - Variation of parameters
   - Euler (-Cauchy) equations
   - Special non-linear ordinary differential equations
   - Series solutions of ordinary differential equations
   - Laplace transform methods
   - Matrix methods for solving systems of ordinary differential equations

The theory for each topic is reviewed and illustrated by example.

The ten examinations in Part 2 are divided into five *midterm* and five *final* examinations. Specifically, the following topics are examined in each case.

Midterm Examination Topics

1. **Terminology, notation and basic concepts associated with the theory of differential equations.**

2. **Solution Techniques for first order ordinary differential equations:**

   - Separable equations
   - Homogeneous equations
   - Exact equations
   - Solution using integrating factors
   - Linear equations
   - Bernoulli equations
   - General strategy for solving first order equations
   - Applications - orthogonal trajectories

3. **Solution Techniques for Higher Order Differential Equations**

   - Theory of linear ordinary differential equations
   - Linear homogeneous differential equations with constant coefficients
   - Linear inhomogeneous differential equations with constant coefficients : *The method of Undetermined Coefficients*
   - Reduction of order

Final Examination Topics

1. *Selected midterm examination topics,* mainly:

   - Selected solution techniques for first order equations
   - The method of undetermined coefficients
   - Reduction of order

2. **Solution Techniques for Higher Order Differential Equations**

- Variation of parameters
- Euler (-Cauchy) equations
- Special non-linear ordinary differential equations
- Series solutions of ordinary differential equations
- Laplace transform methods
- Matrix methods for solving systems of ordinary differential equations

The problems presented in the examinations are representative of the material in most introductory courses in ordinary differential equations. It should be noted, however, that an individual instructor may choose to emphasize certain topics more than others (and may also include or exclude some) in any particular examination. For this reason, students should stay in touch with the individual instructor's requirements.

Working with the Examinations

(a) The examinations are rated on a scale of 1 to 5. Only levels 3 to 5 are addressed in this text. Levels 3 and 4 represent the typical standard while level 5 is more challenging. Start with levels 3 and 4 and move onto level 5 when you have gained the necessary confidence.

(b) All examinations include an allotted time. This will give you some idea of the rate at which you should be completing the problems. Don't be too concerned about this. The point here is to expose yourself to as *many typical examination problems* (and their solutions!) as possible (i.e. this is a learning exercise). With this in mind, try to struggle a little before consulting the solutions. Later, when you have sufficient confidence, you might try any one of the examinations under *actual examination conditions* - you can even grade them yourself.

(c) Calculators are not necessary for any of the problems.

(d) Students who have two midterms can choose, in addition to the midterm examinations in Part 2, specific questions from the final examinations, given the individual instructor's requirements.

(e) At final examination time, don't forget to use selected midterm examination questions for practice. Remember, the final examination usually includes questions examining topics offered before the midterm examination.

(f) Since calculus is such a vital component in the solution of differential equations, the Appendix includes most of the relevant techniques from integral and differential calculus. There are also tables of derivatives, integrals and certain supplementary formulas.

Using the Solutions

The solutions are written more to provide teaching assistance than to furnish a set of answers. Many of the solutions are detailed and include the relevant theory or formulae (from Part 1) used in arriving at the correct answer. Study the solutions carefully. Try to mimic the *procedure.* Differential equations are solved mainly by identifying a specific *class* to which a particular differential equation belongs. Associated with each class, there is a specific *solution procedure*: a sequence of logical steps used to solve each equation in that class. Once the class has been *identified*, the appropriate *procedure* is applied to solve the differential equation. Usually, within a class, the solution of a particular differential equation differs from that of another only in (calculus) detail. This is exactly the approach adopted in the solutions:

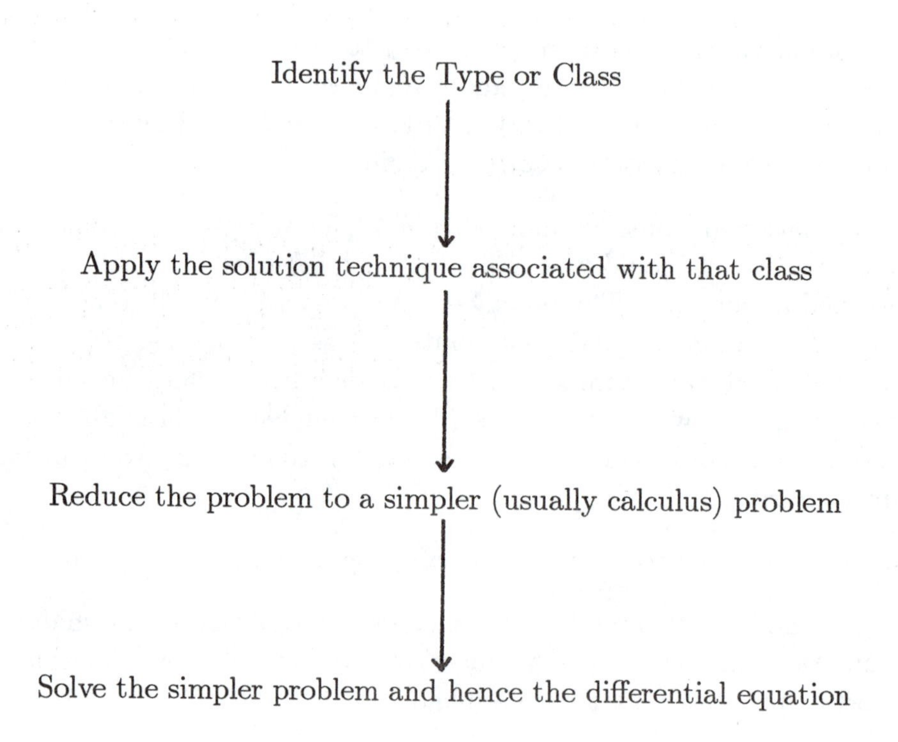

Strategies for Success

Clearly what you do during term-time will significantly affect the outcome of your examination. In mathematics three things stand out above all others:

(a) **Solve as many problems as possible** - For example, assignment questions, questions from the textbook. The more problems you attempt, the more likely you will become *fluent* in the relevant techniques (mathematics is basically a language - you have to practice in order to stay fluent). This will certainly make your examination preparation much easier.

(b) **Ask!** Ask as many questions as you need to ask - but do so politely and professionally. This is an essential part of the learning process.

(c) **Work through practice examinations!** Think of a course examination in the same way that you would a driving test: it is not sufficient to know how to drive - one must also be able to perform certain manoeuvres under the specific constraints of the test. Most people taking a driving test practice *the test* before the test itself. That is, they perform a *rehearsal* or *dress-rehearsal.* The same is true of course examinations. Practice examinations give you the opportunity to *experience* the "event" before the actual examination. This type of "experience" is invaluable. In addition, it adds to confidence and decreases the potential for *exam-anxiety.*

Notations Used Throughout the Text

Standard mathematical notations are employed throughout the text. In addition, any reference to a bracketed number should be interpreted as a reference to the corresponding previously defined equation or formula. For example, the phrase, "From (3.2.1), it is clear that ...." refers to the formula or equation previously labeled as (3.2.1).

PART 1

REVIEW OF SOLUTION TECHNIQUES FOR ORDINARY DIFFERENTIAL EQUATIONS

CHAPTER 1
Basic Concepts

In this chapter, we review some of the basic notation and terminology associated with the theory of differential equations.

Differential Equation

A differential equation is an equation involving an *independent variable*, a *dependent variable* **and** the *derivatives* of the dependent variable. For example

$$\frac{d^3y}{dx^3} - x\frac{dy}{dx} = \sin x \tag{1.1}$$

is a differential equation with independent variable x and dependent variable y (depends on x). On the other hand

$$\frac{d^2x}{dt^2} + x = \cos t - t\frac{dx}{dt} \tag{1.2}$$

is a differential equation with independent variable t and dependent variable x (depends on t). The *dependent variable* (y in (1.1) and x in (1.2)) is also referred to as the *unknown function:* that which is to be found from the solution process.

Differential equations are classified according to various criteria:

Ordinary Differential Equation versus Partial Differential Equation

If the differential equation contains only ordinary (as opposed to partial) derivatives, it is referred to as an *ordinary differential equation.* Otherwise, it is known as a *partial differential equation.* For example, both (1.1) and (1.2) above are ordinary differential equations whereas

$$\frac{\partial^2 z}{\partial x^2} + \frac{\partial^2 z}{\partial y^2} = e^{x+y}$$

is a partial differential equation (it contains partial derivatives of the unknown (dependent) variable z).

The Order of a Differential Equation

The *order* of a differential equation is the order of the highest derivative appearing anywhere in the equation. For example, (1.1) is a *third* order ordinary differential equation whereas (1.2) is a *second* order ordinary differential equation.

The Degree of a Differential Equation

The *degree* of a differential equation is the power to which the differential equation's highest-order derivative has been raised. For example, the differential equation

$$\frac{d^3y}{dx^3} - x\frac{dy}{dx} = 5$$

is of *first degree* (the highest derivative appearing is of the third order and is raised to the power 1) although of third order. On the other hand, the differential equation

$$\left(\frac{d^4y}{dx^4}\right)^5 + \frac{d^3y}{dx^3} - x\left(\frac{dy}{dx}\right)^2 = \sin x \tag{1.3}$$

is of *degree 5* (the highest derivative appearing is of order four and is raised to the power 5) although of fourth order. The equation (1.3) is therefore a fourth order, fifth degree, ordinary differential equation.

Linearity

An nth-order ordinary differential equation in the unknown (dependent) variable y and independent variable x is said to be *linear* if it "fits" the following form:

$$a_n(x)\frac{d^ny}{dx^n} + a_{n-1}(x)\frac{d^{n-1}y}{dx^{n-1}} + \ldots + a_1(x)\frac{dy}{dx} + a_0(x)y = f(x) \tag{1.4}$$

Here, the functions $a_j(x)$ $(j = 0, 1, ..., n)$ and $f(x)$ are known and depend only on the variable x. Notice that y and any of its derivatives are allowed to have only power one and no term can involve products of *different* derivatives of the dependent variable. For example

$$\begin{aligned} \frac{d^3y}{dx^3} - x\frac{dy}{dx} &= 5, \\ \frac{d^2y}{dx^2} - 3\frac{dy}{dx} &= \cos x, \\ \frac{dy}{dx} + y\cos x &= 5 \end{aligned}$$

are all linear, ordinary differential equations (as are (1.1) and (1.2)). Ordinary differential equations which do not fit the form (1.4) are said to be *nonlinear*. For example, (1.3) is a nonlinear ordinary differential equation, as is the equation

$$\frac{d^3y}{dx^3} + \left(\frac{d^2y}{dx^2}\right)\left(\frac{dy}{dx}\right) + y = 0$$

which involves products of *different* derivatives of the dependent variable.

Alternative Notations

Derivatives such as $\frac{d^ny}{dx^n}, ..., \frac{d^3y}{dx^3}, \frac{d^2y}{dx^2}$ and $\frac{dy}{dx}$ are often represented using the 'prime notation', thatis, respectively, $y^{(n)}, ..., y''', y''$ and y'. When the independent variable is t (for time) this notation is altered slightly to a "dot notation"

$$\frac{dy}{dt} = \dot{y}, \quad \frac{d^2y}{dt^2} = \ddot{y} \text{ etc}$$

Hence, (1.1) and (1.2) could alternatively be written as

$$\begin{aligned} y''' - xy' &= \sin x, \\ \ddot{x} + x &= \cos t - t\dot{x} \end{aligned}$$

respectively.

Example 1.1 The differential equation

$$\frac{d^3y}{dx^3} - x\frac{dy}{dx} = 5$$

is a third order, linear, ordinary differential equation.

The differential equation (1.3)

$$\left(\frac{d^4y}{dx^4}\right)^5 + \frac{d^3y}{dx^3} - x\left(\frac{dy}{dx}\right)^2 = \sin x$$

is a fourth-order, fifth-degree, nonlinear, ordinary differential equation.

The Solution of a Differential Equation

A solution of a differential equation is any function (representing the dependent variable) which satisfies (i.e. "fits") the differential equation.

Example 1.2 The functions $y = \sin x$ and $y = \cos x$ are solutions of the linear, second order, ordinary differential equation,

$$y'' + y = 0 \tag{1.5}$$

To see this, note that, with $y = \sin x$

$$\begin{aligned} y'' + y &= \frac{d^2}{dx^2}(\sin x) + \sin x \\ &= -\sin x + \sin x \\ &= 0, \text{ as required.} \end{aligned}$$

Similarly, with $y = \cos x$,

$$\begin{aligned} y'' + y &= \frac{d^2}{dx^2}(\cos x) + \cos x \\ &= -\cos x + \cos x \\ &= 0, \quad \text{as required.} \end{aligned}$$

The collection of *all* solutions of linear, ordinary differential equations (i.e. differential equations of the form (1.4)) is known as the *General Solution.* For example, all solutions of the differential equation (1.5) take the form

$$y = c_1 \cos x + c_2 \sin x, \quad c_1, c_2 \text{ are arbitrary constants} \tag{1.6}$$

which is therefore the general solution of (1.5). The presence of arbitrary constants in (1.6) is not unexpected since the retrieval of a function y from a differential equation inevitably involves at least one integration (or anti-differentiation), the exact number of integrations and hence arbitrary constants in the general solution, depending on the order of the differential equation.

The theory of ordinary differential equations is concerned with the development of systematic methods for *solving* (finding general solutions of) ordinary differential equations.

Initial and Boundary Value Problems

In some cases, the solution of a differential equation is required to satisfy certain additional conditions known as initial (when all conditions occur at the *same* value of the independent variable) or boundary (when conditions occur at *different* values of the independent variable) conditions. For example, if we seek a solution of (1.5) satisfying the conditions $y = 1$ when $x = \frac{\pi}{2}$ and $y = 2$ when $x = \pi$ (also written as $y(\frac{\pi}{2}) = 1$ and $y(\pi) = 2$), then we would pick from the general solution (1.6) the particular solution that satisfyies these additional conditions. In other words, we would use the (boundary) conditions to evaluate the arbitrary constants:

$$\begin{aligned} y &= c_1 \cos x + c_2 \sin x \\ 1 &= c_1 \cos\left(\frac{\pi}{2}\right) + c_2 \sin\left(\frac{\pi}{2}\right) \\ 1 &= c_2 \end{aligned}$$

Similarly,

$$\begin{aligned} y &= c_1 \cos x + \sin x \\ 2 &= c_1 \cos(\pi) + \sin(\pi) \\ 2 &= -c_1 \\ c_1 &= -2 \end{aligned}$$

Finally, the particular or specific solution of interest is given by

$$y = -2\cos x + \sin x$$

Problems requiring the solution of a differential equation subject to initial or boundary conditions are known as initial or boundary value problems, respectively. Note that the number of conditions required to find a particular solution from a general solution is the same as the number of arbitrary constants in the general solution (or the same as the order of the differential equation).

CHAPTER 2
First Order Ordinary Differential Equations

In this chapter, we will be concerned with first order ordinary differential equations of the form

$$\frac{dy}{dx} = f(x, y) \tag{2.1}$$

or (in differential form)

$$dy - f(x, y)dx = 0 \tag{2.2}$$

where $f(x, y)$ is a given function of x and y. Note that (2.1) is not necessarily a linear equation: depending on the specific form of $f(x, y)$, (2.1) may or may not "fit" the form (1.4).

We shall see that the key to identifying the appropriate solution technique for solving (2.1) is very much dependent on the particular form or "shape" of the function $f(x, y)$. Once this is done, the *differential equation problem* reduces to a *calculus problem*, that is, one of integration.

In what follows, c and c_i , $i = 1, 2, 3\ldots$ will denote arbitrary constants of integration.

2.1 Separable Equations

Suppose in (2.1), $f(x, y)$ separates into a product of two functions, one a function of x only and one a function of y only:

$$f(x, y) = h(x)g(y)$$

In this case, (2.1) becomes

$$\frac{dy}{dx} = h(x)g(y) \tag{2.1.1}$$

which is solved by 'gathering together' all the x's on one side and all the y's on the other and then integrating both sides:

$$\begin{aligned} \frac{dy}{g(y)} &= h(x)dx \\ \int \frac{dy}{g(y)} &= \int h(x)dx \end{aligned}$$

This reduces the *differential equation problem* to a problem in *integration.* (Note that either or both of h and g are allowed to be constant functions - in which case the solution of (2.1.1) is found by direct integration.)

Example 2.1.1 Solve the initial value problem

$$\frac{y}{4+y^2}dy = \frac{\sec^2 x}{\tan x}dx, \qquad y(\frac{\pi}{4}) = 0$$

Solution. First write the equation in the form (2.1)

$$\frac{dy}{dx} = \left(\frac{4+y^2}{y}\right)\left(\frac{\sec^2 x}{\tan x}\right) \tag{2.1.2}$$

We note that the right-hand side of (2.1.2) separates into a product of a function of x only and a function of y only. Our equation is of the form (2.1.1), that is, separable. Proceed by "gathering up" the x's on one side and the y's on the other and integrating:

$$\int \frac{y}{4+y^2}dy = \int \frac{\sec^2 x}{\tan x}dx$$

(The problem is now reduced to one of integration). Using the substitution rule and rules for logarithms and exponentials (see the Appendix), we obtain

$$\begin{aligned} \frac{1}{2}\ln\left(4+y^2\right) &= \ln|\tan x| + c \\ \ln\left(4+y^2\right) &= \ln(\tan x)^2 + c_1 \qquad (c_1 = 2c) \\ 4+y^2 &= \left(\tan^2 x\right)\exp(c_1) \\ &= c_2\tan^2 x \qquad (c_2 = \exp(c_1)) \end{aligned}$$

Hence,

$$y^2 = c_2\tan^2 x - 4$$

This is the *general solution* of the differential equation. Finally, apply the condition $y(\frac{\pi}{4}) = 0:$

$$\begin{aligned} 0^2 &= c_2\tan^2\left(\frac{\pi}{4}\right) - 4 \\ 0 &= c_2(1) - 4 \\ c_2 &= 4 \end{aligned}$$

The solution of the boundary value problem (particular solution of the differential equation) is thus given by

$$y^2 = 4(\tan^2 x - 1)$$

2.2 Homogeneous Equations

Consider again the first order ordinary differential equation (2.1).

$$\frac{dy}{dx} = f(x, y)$$

This time, suppose the right-hand side $f(x, y)$ has the following property.

$$f(\lambda x, \lambda y) = f(x, y) \qquad \text{for any real number } \lambda \neq 0 \tag{2.2.1}$$

Then (2.1) is referred to as a *homogeneous* first order differential equation and the substitution

$$y = vx, \qquad \text{where } v \text{ is some unknown function of } x \tag{2.2.2}$$

will reduce the homogeneous equation to a separable equation (which is then solved as in Section 2.1).

Example 2.2.1 Solve the differential equation

$$y' = \frac{x^2 + y^2}{xy} \tag{2.2.3}$$

Solution. Note that (2.2.3) is not separable since the right-hand side $\frac{x^2 + y^2}{xy}$ cannot be decomposed into a product of a function of x only and a function of y only. To see if (2.2.3) is homogeneous, we attempt to verify (2.2.1) by writing $f(x, y) = \frac{x^2 + y^2}{xy}$ and replacing x by λx and y by λy:

$$\begin{aligned} f(\lambda x, \lambda y) &= \frac{(\lambda x)^2 + (\lambda y)^2}{(\lambda x)(\lambda y)} \\ &= \frac{\lambda^2 x^2 + \lambda^2 y^2}{\lambda^2 xy} \\ &= \frac{x^2 + y^2}{xy} \\ &= f(x, y) \end{aligned}$$

Hence, (2.2.3) is indeed homogeneous so that the substitution $y = vx$ will reduce (2.2.3) to a separable equation.

Let $y = vx$, then, using the product rule,

$$\frac{dy}{dx} = v + x\frac{dv}{dx} \tag{2.2.4}$$

Substituting (2.2.2) and (2.2.4) into (2.2.3), we obtain

$$\begin{aligned} v + x\frac{dv}{dx} &= \frac{x^2 + (vx)^2}{x\,(vx)} \\ &= \frac{x^2 + v^2x^2}{vx^2} \\ &= \frac{1+v^2}{v} = \frac{1}{v} + v \\ \frac{dv}{dx} &= \frac{1}{xv} = \left(\frac{1}{x}\right)\left(\frac{1}{v}\right), \qquad \text{which is separable} \end{aligned}$$

As in Section 2.1,

$$\begin{aligned} \int v dv &= \int \frac{dx}{x} \\ \frac{v^2}{2} &= \ln|x| + c \end{aligned}$$

Now let $v = \dfrac{y}{x}$,

$$\begin{aligned} \frac{y^2}{x^2} &= 2\ln|x| + c_1 \qquad (c_1 = 2c) \\ y^2 &= x^2 \ln x^2 + c_1 x^2 \end{aligned}$$

Note 2.2.2 The substitution $x = vy$ will also reduce a homogeneous equation to a separable equation. This substitution is sometimes preferable to (2.2.2) when (2.1) is written in the form

$$\frac{dy}{dx} = f(x, y) = -\frac{M(x, y)}{N(x, y)}$$

or

$$M(x, y)dx + N(x, y)dy = 0$$

and $M(x, y)$ is "simpler" than $N(x, y)$.

2.3 Exact Equations

Suppose the right-hand side of (2.1) is written in the form

$$f(x, y) = -\frac{M(x, y)}{N(x, y)}, \qquad M, N \text{ are functions of } x, y$$

so that (2.1) becomes

$$M(x, y)dx + N(x, y)dy = 0 \qquad (2.3.1)$$

If it is possible to find a function $F(x, y)$ such that

$$\frac{\partial F}{\partial x} = M \quad \text{and} \quad \frac{\partial F}{\partial y} = N \tag{2.3.2}$$

then (2.3.1) becomes

$$\frac{\partial F}{\partial x} dx + \frac{\partial F}{\partial y} dy = dF = 0 \tag{2.3.3}$$

which has the simple solution

$$F(x, y) = c = \text{ constant} \tag{2.3.4}$$

In fact, it can be shown that such an F (satisfying (2.3.2)) will exist for "suitable" functions M and N if and only if

$$\frac{\partial M}{\partial y} = \frac{\partial N}{\partial x} \tag{2.3.5}$$

When (2.3.3) is true, the ordinary differential equation (2.3.1) is said to be *exact* (since the left-hand side of (2.3.3) is an exact differential).

Procedure for Solving Exact Equations

1. Identify M and N by writing the differential equation in the form (2.3.1).
2. Verify (2.3.5).
3. If (2.3.5) is true, find F using (2.3.2) (see Example 2.3.1 below).
4. The general solution of the differential equation is given by (2.3.4).

Example 2.3.1 Solve

$$\frac{dy}{dx} = \frac{3x^2 \sin^2 y}{2e^{2y} - 2x^3 \sin y \cos y}$$

Solution. It is clear from the right-hand side of the equation that it is neither separable nor homogeneous. We try for 'exactness' by first identifying M and N. Write the differential equation in the form (2.3.1).

$$\underbrace{3x^2 \sin^2 y}_{M}\, dx + \underbrace{(2x^3 \sin y \cos y - 2e^{2y})}_{N}\, dy = 0$$

Next, apply the test for exactness (2.3.5):

$$\frac{\partial M}{\partial y} = 6x^2 \sin y \cos y = \frac{\partial N}{\partial x}$$

Hence, the general solution of the differential equation is given by

$$F = \text{ constant}$$

where F is given by (2.3.2):

$$\frac{\partial F}{\partial x} = M = 3x^2 \sin^2 y \tag{2.3.6a}$$

and

$$\frac{\partial F}{\partial y} = N = 2x^3 \sin y \cos y - 2e^{2y} \tag{2.3.6b}$$

Choose either of (2.3.6a), (2.3.6b) and integrate. From (2.3.6a)

$$F(x, y) = x^3 \sin^2 y + g(y), \qquad g \text{ is an arbitrary function of } y \text{ only} \tag{2.3.7}$$

We find the unknown function g by imposing (2.3.6b) (the other requirement for F) on (2.3.7). To do this, note from (2.3.7)

$$\frac{\partial F}{\partial y} = 2x^3 \sin y \cos y + g'(y)$$

If this expression is to be the same as (2.3.6b), we require that

$$\begin{aligned} g'(y) &= -2e^{2y} \\ g(y) &= -e^{2y} + c_1 \end{aligned}$$

Finally, from (2.3.7)

$$F(x, y) = x^3 \sin^2 y - e^{2y} + c_1$$

From (2.3.4), the general solution of the differential equation is given by

$$F(x, y) = x^3 \sin^2 y - e^{2y} + c_1 = c$$

or

$$x^3 \sin^2 y - e^{2y} = c_2 \,, \qquad c_2 = c - c_1$$

Note 2.3.2 If we start with (2.3.6b) and impose (2.3.6a) (instead of the other way around), we obtain exactly the same general solution.

2.4 Integrating Factors

Sometimes, when a first order ordinary differential equation is not exact, we can make it so by multiplying both sides of the equation by a suitable *integrating factor*. The resulting equation is then solved using the method discussed in Section 2.3. The procedure for finding and using an integrating factor is as follows.

1. Write the ordinary differential equation in the form (2.3.1).

2. Examine the quantity

$$\frac{1}{N}\left(\frac{\partial M}{\partial y}-\frac{\partial N}{\partial x}\right) \tag{2.4.1}$$

If (2.4.1) is a function of x only, that is, $\frac{1}{N}\left(\frac{\partial M}{\partial y}-\frac{\partial N}{\partial x}\right)=R(x)$, say, then the differential equation (2.3.1) has an integrating factor given by

$$\mu(x)=\exp\left(\int R(x)dx\right)=\exp\left(\int \frac{1}{N}\left(\frac{\partial M}{\partial y}-\frac{\partial N}{\partial x}\right)dx\right) \tag{2.4.2}$$

Now skip to Step #4.

3. If (2.4.1) is not a function only of x, examine instead the quantity

$$\frac{1}{M}\left(\frac{\partial N}{\partial x}-\frac{\partial M}{\partial y}\right) \tag{2.4.3}$$

If (2.4.3) is a function of y only, that is, $\frac{1}{M}\left(\frac{\partial N}{\partial x}-\frac{\partial M}{\partial y}\right)=H(y)$, say, then the differential equation (2.3.1) has an integrating factor given by

$$\mu(y)=\exp\left(\int H(y)dy\right)=\exp\left(\int \frac{1}{M}\left(\frac{\partial N}{\partial x}-\frac{\partial M}{\partial y}\right)dy\right) \tag{2.4.4}$$

4. Multiply both sides of the ordinary differential equation (2.3.1) by either $\mu(x)$ or $\mu(y)$ (as appropriate). The equation is now exact and can be solved using the methods of Section 2.3.

Example 2.4.1 Solve

$$\frac{dy}{dx}=-\frac{2\sin\left(y^2\right)}{xy\cos(y^2)}$$

Solution. Following the above procedure,

1. We first write the differential equation in the form (2.3.1).

$$\underbrace{2\sin\left(y^2\right)}_{M}dx+\underbrace{xy\cos\left(y^2\right)}_{N}dy=0$$

2. Next, we determine the required partial derivatives with each of (2.4.1) and (2.4.3) in mind.

$$\frac{\partial M}{\partial y}=4y\cos\left(y^2\right); \qquad \frac{\partial N}{\partial x}=y\cos\left(y^2\right)$$

Examine (2.4.1).

$$\begin{aligned}\frac{1}{N}\left(\frac{\partial M}{\partial y}-\frac{\partial N}{\partial x}\right) &= \frac{1}{xy\cos\left(y^2\right)}\left(3y\cos\left(y^2\right)\right)\\ &= \frac{3}{x}\\ &= R(x)\end{aligned}$$

Clearly, (2.4.1) is indeed a function of x only! Hence, an integrating factor $\mu(x)$ can be found from (2.4.2). (Note that there is no need to examine (2.4.3) i.e. we can skip Step #3). In fact, from (2.4.2)

$$\begin{aligned}\mu(x) &= \exp(\int \frac{3}{x}dx)\\ &= \exp(\ln|x|^3+c)\\ &= c_1|x|^3 \qquad (c_1=e^c)\end{aligned}$$

We need only one of these integrating factors (they will all do the same job) so we choose

$$\mu(x)=x^3$$

4. Next, we multiply both sides of our differential equation by $\mu(x)=x^3$.

$$2x^3\sin\left(y^2\right)dx+x^4y\cos\left(y^2\right)dy=0$$

This equation is now exact (in the sense of Section 2.3) and its general solution can be found as in Section 2.3:

$$x^4\sin\left(y^2\right)=c$$

2.5 Linear First Order Differential Equations

Consider the first order (linear) differential equation from (1.4)

$$\frac{dy}{dx}+P(x)y=Q(x) \tag{2.5.1}$$

Here, P and Q are given functions of x. This equation is a special case of the class studied in Section 2.4. In fact, an integrating factor for (2.5.1) is given by

$$\mu(x)=\exp\left(\int P(x)dx\right) \tag{2.5.2}$$

To solve (linear) equations of the form (2.5.1), we first find the integrating factor (2.5.2) and then multiply both sides of (2.5.1) by $\mu(x)$

$$\mu(x)\left(\frac{dy}{dx}+P(x)y\right)=\mu(x)Q(x)$$

This equation is now exact - but rather than solve it using the methods of Section 2.3, we note that it can also be can be written in the form

$$\frac{d}{dx}(\mu(x)y) = \mu(x)Q(x) \tag{2.5.3}$$

which is *separable* and can be solved as in Section 2.1.

Example 2.5.1 Solve

$$\frac{dy}{dx} = x + y$$

Solution. We first note that the equation is of the form (2.5.1):

$$\frac{dy}{dx} \underbrace{-\ y}_{P(x)} = \underbrace{x}_{Q(x)}$$

Consequently, an integrating factor is given by (2.5.2).

$$\begin{aligned} \mu(x) &= \exp\left(\int -1dx\right) \\ &= \exp(-x + c) \end{aligned}$$

Choose

$$\mu(x) = e^{-x}$$

Multiply both sides of the differential equation by $\mu(x)$.

$$e^{-x}\left(\frac{dy}{dx} - y\right) = xe^{-x}$$

From (2.5.3), this equation can be written in the form

$$\frac{d}{dx}\left(ye^{-x}\right) = xe^{-x} \tag{2.5.4}$$

(This is easily verified by expanding the left-hand side of (2.5.4) using the rule for differentiating products of functions (see the Appendix)). Since (2.5.4) is now separable, we can write (see Section 2.1)

$$\begin{aligned} \int d\left(ye^{-x}\right) &= \int xe^{-x}dx \\ ye^{-x} &= \int xe^{-x}dx \end{aligned}$$

The remaining integral is found using integration by parts (Appendix).

$$ye^{-x} = -e^{-x}(1 + x) + c$$

Finally, the general solution of the differential equation is given by

$$y = -(1+x) + ce^x$$

Note 2.5.2 Sometimes the differential equation is not *linear in y* but *linear in x*. Consider the differential equation

$$\frac{dy}{dx} = \frac{y}{y^2 - 2x}$$

The equation does not (immediately) fit the form (2.5.1) so cannot be solved as in Example 2.5.1. However, if we rewrite the equation as

$$\begin{aligned}\frac{dx}{dy} &= \frac{y^2 - 2x}{y} \\ \frac{dx}{dy} + \frac{2}{y}x &= y\end{aligned}$$

we note that the equation is indeed of the form (2.5.1) with y and x switched. That is, the differential equation is not linear in y but *is linear in x*! Consequently, we can apply the procedure discussed above but with x and y switched. An integrating factor is given by

$$\begin{aligned}\mu(y) &= \exp\left(\int \frac{2}{y}dy\right) \\ &= \exp(\ln y^2 + c)\end{aligned}$$

Choose

$$\mu(y) = y^2$$

Consequently,

$$\begin{aligned}y^2\left(\frac{dx}{dy} + \frac{2}{y}x\right) &= y^3 \\ \frac{d}{dy}\left(xy^2\right) &= y^3\end{aligned}$$

which is again separable. In fact,

$$\begin{aligned}\int d\left(xy^2\right) &= \int y^3 dy \\ xy^2 &= \frac{y^4}{4} + c\end{aligned}$$

so that the general solution is given by

$$x = \frac{y^2}{4} + \frac{c}{y^2}$$

2.6 Bernoulli Equations

The Bernoulli equation

$$\frac{dy}{dx} + P(x)y = Q(x)y^n, \qquad n \in R \tag{2.6.1}$$

is almost of the form (2.5.1) except for the nonlinear term in y^n on the right-hand side (note that the case $n = 0$ is exactly the case discussed in Section 2.5 while $n = 1$ corresponds to a separable equation (§2.1)).

The substitution

$$v(x) = y^{1-n} \tag{2.6.2}$$

will reduce the equation (2.6.1) to the (linear in v) equation

$$\left(\frac{1}{1-n}\right)\frac{dv}{dx} + P(x)v = Q(x) \tag{2.6.3}$$

which can be solved using the method of Section 2.5.

Example 2.6.1 Solve

$$y' = x^3y^2 + xy \tag{2.6.4}$$

Solution Comparing the differential equation with (2.6.1), we see that (2.6.4) is of Bernoulli type with

$$P(x) = -x, \qquad Q(x) = x^3, \qquad n = 2$$

According to (2.6.2), let

$$v(x) = y^{1-2} = y^{-1} \tag{2.6.5}$$

so that

$$\begin{aligned} \frac{dv}{dx} &= -y^{-2}\frac{dy}{dx} \\ &= -v^2\frac{dy}{dx} \end{aligned}$$

The differential equation (2.6.4) now becomes

$$-\frac{1}{v^2}\frac{dv}{dx} = \frac{x^3}{v^2} + \frac{x}{v}$$

or, as in (2.6.3),

$$\frac{dv}{dx} + xv = -x^3 \tag{2.6.6}$$

which is of the form (2.5.1), that is, linear in v.

Using the method discussed in Section 2.5, we obtain the general solution of (2.6.6) as

$$v(x) = -x^2 + 2 + c_1 e^{-\frac{1}{2}x^2}$$

Finally, from (2.6.5), since $y = \frac{1}{v}$, the general solution of (2.6.4) is given by

$$\begin{aligned} \frac{1}{y(x)} &= -x^2 + 2 + c_1 e^{-\frac{1}{2}x^2} \\ y(x) &= \frac{1}{-x^2 + 2 + c_1 e^{-\frac{1}{2}x^2}} \\ &= \frac{e^{\frac{1}{2}x^2}}{c_1 + 2e^{\frac{1}{2}x^2} - x^2 e^{\frac{1}{2}x^2}} \end{aligned}$$

2.7 General Strategy for Solving First Order ODEs

The key to solving a first order ordinary differential equation is in *classification*: identifying the equation as belonging to a certain *class* (separable, homogeneous, exact etc) and using the technique for that *class* to solve the equation. The most difficult part of the solution process is invariably the *classification* or *identification* of the equation. Once this is accomplished, the remainder is basically calculus.

2.8 Orthogonal Trajectories

Suppose we have a family of curves represented by the equation

$$g(x, y, a) = 0 \tag{2.8.1}$$

where a is some parameter (i.e. one curve for each value of a). It is often desirable to identify the corresponding family of curves (with parameter b) which intersect the curves (2.8.1) at right angles (whenever they do intersect). That is, to identify the family

$$h(x, y, b) = 0 \tag{2.8.2}$$

such that whenever a curve from (2.8.2) intersects one from (2.8.1), the corresponding tangents to each curve at the intersection point are orthogonal. To do this, we first find the differential equation corresponding to the family (2.8.1). For example,

$$\frac{dy}{dx} = f(x, y) \tag{2.8.3}$$

The orthogonal trajectories are then given by the solutions of the differential equation

$$\frac{dy}{dx} = -\frac{1}{f(x, y)} \tag{2.8.4}$$

arising from the fact that at each point of intersection, the slopes of the curves (2.8.3) and (2.8.4) (or of the corresponding tangents to the curves) must be negative reciprocals of one another.

Example 2.8.1 Consider the family of circles (with parameter a) :

$$x^2 + y^2 = a^2 \tag{2.8.5}$$

The associated differential equation is found by differentiating (2.8.5) implicitly with respect to x :

$$\begin{aligned} 2x + 2y\frac{dy}{dx} &= 0 \\ \frac{dy}{dx} &= -\frac{x}{y} \end{aligned}$$

The differential equation for the orthogonal trajectories is then given by (from (2.8.4))

$$\frac{dy}{dx} = \frac{y}{x}$$

Solving this (separable - see Section 2.1) differential equation leads to $y(x) = bx$ where b is an arbitrary constant. It follows that the orthogonal trajectories of the family (2.8.5) are given by

$$y = bx, \qquad b \text{ is a parameter.} \tag{2.8.6}$$

In other words, whenever the curves (2.8.5) and (2.8.6) intersect, they will do so in such a way that their tangents (at the intersection points) will be orthogonal to each other.

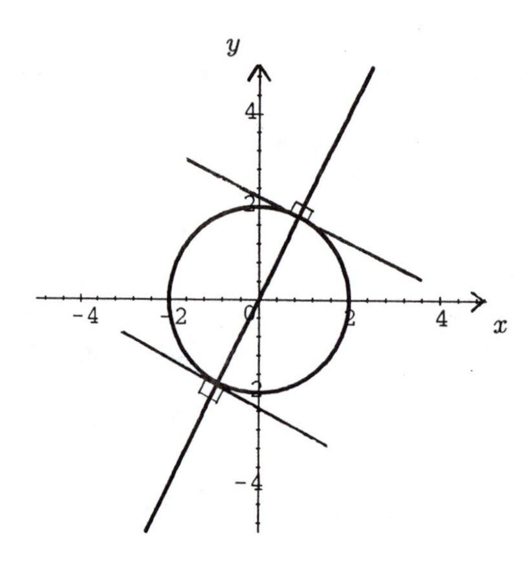

$x^2 + y^2 = 2$ and $y = 2x$

CHAPTER 3
Higher Order Ordinary Differential Equations

In this chapter, we will be concerned with different types of higher order ordinary differential equations but mainly *linear* ordinary differential equations of the form (1.4). Again, $c, c_1, c_2, ...c_n$ will denote arbitrary constants.

3.1 Linear Ordinary Differential Equations

The linear ordinary differential equation (1.4) is amenable to exact solution since its general solution decomposes into two simpler parts, that is, the general solution of (1.4) can be written as

$$y(x) = y_c(x) + y_p(x) \tag{3.1.1}$$

In (3.1.1), $y_c(x)$, referred to as the *complementary solution*, is the general solution of the homogeneous equation associated with (1.4) (the use of the term "homogeneous" in this context is completely different from that in Section 2.2), that is, $y_c(x)$ solves (1.4) with $f(x) \equiv 0$:

$$a_n(x)\frac{d^n y_c}{dx^n} + a_{n-1}(x)\frac{d^{n-1} y_c}{dx^{n-1}} + \ldots + a_1(x)\frac{dy_c}{dx} + a_0(x)y_c = 0 \tag{3.1.2}$$

In fact, y_c always takes the form

$$y_c(x) = \sum_{i=1}^{n} c_i y_i(x) \tag{3.1.3}$$

where $y_i, i = 1, ...n$ are n *linearly independent* solutions of (3.1.2).

Note 3.1.1

(a) A sufficiently differentiable set of functions $f_i(x)$ (with n^{th}-derivative denoted by $f_i^{(n)}(x)$) $i = 1, ..., n$, defined on some interval $I = (a, b)$ is *linearly independent* on I if the Wronskian

$$W(x) = \begin{vmatrix} f_1 & f_2 & \cdots & f_n \\ f_1' & f_2' & \cdots & f_n' \\ \vdots & \vdots & \vdots & \vdots \\ f_1^{(n-1)} & f_2^{(n-1)} & \cdots & f_n^{(n-1)} \end{vmatrix}$$

is nonzero at at least one point in the interval I. (The converse of this statement is not true in general - but *is* true if the n functions are solutions of a homogeneous differential equation - see (b) below)

(b) Let the homogeneous differential equation (3.1.2) with continuous coefficients defined on the interval I have the n solutions $y_i(x)$, $i = 1, ..., n$. Then the *Wronskian*

$$W(x) = \begin{vmatrix} y_1 & y_2 & \cdots & y_n \\ y_1' & y_2' & \cdots & y_n' \\ \vdots & \vdots & \vdots & \vdots \\ y_1^{(n-1)} & y_2^{(n-1)} & \cdots & y_n^{(n-1)} \end{vmatrix} \qquad (3.1.4)$$

is either identically zero or never zero on I. Furthermore, $W(x)$ is identically zero if and only if $y_i(x)$, $i = 1, ..., n$ are *linearly dependent.* Thus

(i) If $W(x) \neq 0$ at at least one point of I, the solutions $y_i(x)$, $i = 1, ..., n$ are *linearly independent* on I.

(ii) If the solutions $y_i(x)$, $i = 1, ..., n$ are *linearly independent,* $W(x) \neq 0$ for *every* $x \in I$.

The term $y_p(x)$ of (3.1.1) is any *particular (known or readily available) solution* of the inhomogeneous equation (1.4) (when $f(x) \neq 0$).

Example 3.1.2 Consider the differential equation

$$y'' + y = x \qquad (3.1.5)$$

The general solution of (3.1.5) is given by (3.1.1) where y_c is the general solution of the homogeneous equation

$$y_c'' + y_c = 0 \qquad (3.1.6)$$

and y_p is any particular solution of the inhomogeneous equation

$$y_p'' + y_p = x \qquad (3.1.7)$$

We shall see in Example 3.3.4 that

$$y_c = c_1 \sin x + c_2 \cos x$$

(in fact, it is easy to verify that $y_1 = \sin x$ and $y_2 = \cos x$ are solutions of (3.1.6). The fact that they are linearly independent follows from Note 3.1.1(b)(i) since

$$\begin{aligned} W(x) &= \begin{vmatrix} \sin x & \cos x \\ \cos x & -\sin x \end{vmatrix} \\ &= -\sin^2 x - \cos^2 x \\ &= -1 \\ &\neq 0 \quad \text{ever!)} \end{aligned}$$

Hence, from (3.1.3) with $n = 2$, it is clear that $y_c = c_1 \sin x + c_2 \cos x$. Also, it can be shown (§3.3 or just guess!) that

$$y_p(x) = x$$

(this is easy to verify by direct substitution into (3.1.7)) so that the general solution of (3.1.5) is given by

$$\begin{aligned} y(x) &= y_c(x) + y_p(x) \\ &= c_1 \sin x + c_2 \cos x + x \end{aligned}$$

From (3.1.1), it is clear that to solve the linear ordinary differential equation (1.4), it is necessary to develop systematic methods for finding both y_c and y_p. This is extremely difficult in general but relatively simple when the coefficients a_i, $i = 0, 1, ..., n$ in (1.4) are *constant.* This special case will be the subject of the next two sections of this chapter.

3.2 Linear Homogeneous Differential Equations with Constant Coefficients

In this section, we discuss a method for obtaining $y_c(x)$, the general solution of the homogeneous equation (3.1.2), in the particular case when the coefficients a_i, $i = 0, 1, ..., n$ are *constant.*

Consider then the linear homogeneous ordinary differential equation from (1.4)

$$a_n(x)\frac{d^n y}{dx^n} + a_{n-1}(x)\frac{d^{n-1}y}{dx^{n-1}} + \ldots + a_1(x)\frac{dy}{dx} + a_0(x)y = 0 \tag{3.2.1}$$

Suppose the coefficients a_i, $i = 0, 1, ..., n$ in (3.2.1) are *constant.* Associated with the differential equation (3.2.1) is a polynomial equation obtained by replacing each of the derivatives $\frac{d^n y}{dx^n}$, $n = 0, 1, 2, ...$ in (3.2.1) by the powers m^n of a new variable m. That is, the associated or *characteristic equation of (3.2.1)* is given by

$$P(m) = a_n m^n + a_{n-1} m^{n-1} + ... + a_1 m + a_0 = 0 \tag{3.2.2}$$

The form of the general solution of (3.2.1) depends on the nature of the solutions of (3.2.2). The different cases are summarized in the following table.

Table 3.2.1 General Solution of (3.2.1)

$P(m)$	$y(x)$
$(m-m_1)(m-m_2)...(m-m_n)$, $m_1 \neq m_2 \neq ... \neq m_n \in \Re$	$c_1 e^{m_1 x} + c_2 e^{m_2 x} + ... + c_n e^{m_n x}$
$(m-m_1)^n$, $m_1 \in \Re$, $n \in N$	$\left(c_1 + c_2 x + ... c_n x^{n-1}\right) e^{m_1 x}$
$[m-(a+ib)]\,[m-(a-ib)] = (m-a)^2 + b^2$	$(c_1 \cos bx + c_2 \sin bx)\, e^{ax}$
$\left[(m-a)^2 + b^2\right]^n$, $a, b \in \Re$, $n \in N$	$\left(c_1 + c_2 x + ... c_n x^{n-1}\right) e^{ax} \cos bx$ $+ \left(d_1 + d_2 x + ... d_n x^{n-1}\right) e^{ax} \sin bx$

Here, $d_1, d_2, ...d_n$ are arbitrary constants. For $P(m)$ given by a product of any of the above forms in the left-hand side of Table 3.2.1, we simply add the corresponding forms from the right-hand side of Table 3.2.1 to form the general solution of (3.2.1).

Example 3.2.2 Solve

$$y''' - 6y'' + 11y' - 6y = 0$$

Solution. First write down the associated characteristic equation.

$$P(m) = m^3 - 6m^2 + 11m - 6 = 0$$

Factor the left-hand side of this equation.

$$\begin{aligned} (m-3)(m-2)(m-1) &= 0 \\ m &= 3, 2, 1 \\ m_1 &= 1, \quad m_2 = 2, \quad m_3 = 3 \end{aligned}$$

From Table 3.2.1, the general solution of the differential equation is given by

$$y(x) = c_1 e^{x} + c_2 e^{2x} + c_3 e^{3x}$$

(Hence, the y_c for the associated inhomogeneous equation

$$y''' - 6y'' + 11y' - 6y = f(x)$$

is given by $y_c(x) = c_1 e^{x} + c_2 e^{2x} + c_3 e^{3x}$.)

Example 3.2.3 Find the complementary solution for the equation

$$y'' - 2y' + y = e^x \tag{3.2.3}$$

Solution. The complementary solution y_c is the general solution of the homogeneous equation

$$y'' - 2y' + y = 0 \tag{3.2.4}$$

To find y_c,first write down the associated characteristic equation.

$$P(m) = m^2 - 2m + 1 = 0$$

Factor the left-hand side of this equation.

$$\begin{aligned} (m-1)^2 &= 0 \\ m &= 1 \ \text{(twice)} \end{aligned}$$

From Table 3.2.1 (with $n = 2$), the general solution of the homogeneous differential equation (3.2.4) is given by

$$y(x) = (c_1 + c_2 x)e^x$$

Hence, the y_c for the associated inhomogeneous equation (3.2.3) is given by

$$y_c(x) = (c_1 + c_2 x)e^x$$

Example 3.2.4 Solve

$$y''' + y'' + y' + y = 0$$

Solution. First write down the associated characteristic equation.

$$P(m) = m^3 + m^2 + m + 1 = 0$$

Factor the left-hand side of this equation.

$$\begin{aligned} (m+1)\left(m^2+1\right) &= 0 \\ m &= -1, \pm i \end{aligned}$$

From Table 3.2.1 (with $a = 0$, $b = 1$), the general solution of the differential equation is given by

$$y(x) = \underbrace{c_1 e^{-x}}_{\text{from } m=-1} + \underbrace{c_2 \cos x + c_3 \sin x}_{\text{from } m=\pm i}$$

(Hence, the y_c for the associated inhomogeneous equation

$$y''' + y'' + y' + y = f(x)$$

is given by $y_c(x) = c_1 e^{-x} + c_2 \cos x + c_3 \sin x$.)

3.3 Linear Inhomogeneous Differential Equations with Constant Coefficients

In this section, we discuss a method for obtaining y_p, a particular solution of the inhomogeneous equation (1.4) ($f(x) \neq 0$). There are several different ways of finding y_p. However, when the coefficients a_i, $i = 0, 1, ..., n$ of the differential equation are *constant*, one of the most systematic ways of finding y_p is the *method of undetermined coefficients.* This method uses the form of the function $f(x)$ on the right-hand side of the differential equation (1.4) to suggest a form or "shape" for $y_p(x)$. The latter contains undetermined constants which are evaluated by substituting the suggested $y_p(x)$ back into the differential equation. Although convenient and easy to use, the method of undetermined coefficients is limited in that it can be used only when $f(x)$ takes on one of the specific forms in Table 3.3.1 below. For any other $f(x)$, y_p must be found using some other technique (e.g. *Variation of Parameters* - see Section 3.5).

In the following table, "suggestions" for y_p are listed alongside the corresponding different forms of the given function $f(x)$.

Table 3.3.1 Particular Solutions

Term in $f(x)$	Corresponding suggestion for term in $y_p(x)$
C	$x^s A$
$Ce^{\beta x}$	$x^s A e^{\beta x}$
$C_1 \sin mx + C_2 \cos mx$	$x^s (A_1 \sin mx + A_2 \cos mx)$
$R_n(x)$	$x^s P_n(x)$
$R_n(x) e^{\beta x}$	$x^s P_n(x) e^{\beta x}$
$R_n(x) e^{\beta x} \sin mx$	$x^s e^{\beta x} [P_n(x) \cos mx + Q_n(x) \sin mx]$
$R_n(x) e^{\beta x} \cos mx$	$x^s e^{\beta x} [P_n(x) \cos mx + Q_n(x) \sin mx]$

Here, C is a *given* non-zero constant; C_1 and C_2 are *given* constants (not both zero); n is a *given* integer, β and m are *given* real numbers; the polynomials R_n, P_n and Q_n are defined by

$$\begin{aligned} R_n(x) &= a_0 + a_1 x + ... + a_{n-1} x^{n-1} + a_n x^n \\ P_n(x) &= B_0 + B_1 x + ... + B_{n-1} x^{n-1} + B_n x^n \\ Q_n(x) &= D_0 + D_1 x + ... + D_{n-1} x^{n-1} + D_n x^n \end{aligned}$$

where $a_0, a_1, ...a_n$ are *given* constants (some of which may be zero) and $B_0, B_1, ...B_n, D_0, D_1, ...D_n$ are specific constants *to be determined.* Finally, A, A_1 and A_2 are specific constants *to be determined.* The exponent s is the smallest non-negative integer $(0, 1, 2, ...)$ required to ensure that no term in the suggestion for y_p also appears in y_c, that is, no term in the suggestion for y_p should solve the corresponding homogeneous differential equation.

Note 3.3.2 When the given function $f(x)$ contains more than one term, each of which appears in Table 3.3.1, we form the y_p by adding the corresponding suggestions (from Table 3.3.1) for each term in $f(x)$:

$$\begin{aligned} f(x) &= \underbrace{f_1(x) + f_2(x) + \dots}_{\text{Each on left-hand side of Table 3.3.1}} . \\ y_p(x) &= \underbrace{y_{p1}(x) + y_{p2}(x) + \dots}_{\text{Each on right-hand side of Table 3.3.1}} \end{aligned}$$

Example 3.3.3 Solve

$$y'' - 6y' = 6e^{6x} \tag{3.3.1}$$

Solution. Since the equation is linear (of the form (1.4)) we can write the general solution in the form (see (3.1.1))

$$y(x) = y_c(x) + y_p(x)$$

Complementary Solution y_c

The characteristic equation (see (3.2.2)) is given by

$$\begin{aligned} m^2 - 6m &= 0 \\ m(m-6) &= 0 \\ m &= 0, 6 \\ m_1 &= 0, \quad m_2 = 6 \end{aligned}$$

From Table 3.2.1,

$$\begin{aligned} y_c(x) &= c_1 e^{0x} + c_2 e^{6x} \\ &= c_1 + c_2 e^{6x} \end{aligned} \tag{3.3.2}$$

Particular Solution y_p

From (3.3.1), $f(x) = 6e^{6x}$. Hence, from Table 3.3.1,

$$y_p(x) = x^s A e^{6x}$$

To determine the required value of s, start with the lowest possible value of s: $s = 0$ and see if the resulting y_p, that is, $y_p(x) = x^0 A e^{6x} = Ae^{6x}$ solves the corresponding homogeneous differential equation. This can be done by either by substituting the expression $y_p(x) = Ae^{6x}$ directly into the homogeneous differential equation (tedious) or by examining the expression for $y_c(x)$, (3.3.2). The latter is quicker and is based on the fact that y_c "contains" all solutions of the homogeneous differential equation, that is, any solution of the homogeneous differential equation can be obtained from y_c by a particular choice of the constants c_1 and c_2. In fact, in this case, if we let $c_1 = 0$ and $c_2 = A$ in (3.3.2) we find that $y_p(x) = Ae^{6x}$ is indeed a solution of the homogeneous differential equation for *any* value of the constant A. Hence, 0 is *not* the correct value for s. Next try $s = 1$:

$$y_p(x) = xAe^{6x} \tag{3.3.3}$$

We see that *no* choice of the *constants* c_1 and c_2 in (3.3.2) will lead to a solution of the form (3.3.3). Consequently, $y_p(x) = xAe^{6x}$ is the correct form for $y_p(x)$.

To find the constant A, we substitute this y_p (from (3.3.3)) back into the inhomogeneous differential equation (3.3.1).

$$\begin{aligned} y_p'' - 6y_p' &= 6e^{6x} \\ \frac{d^2}{dx^2}\left(Axe^{6x}\right) - 6\frac{d}{dx}\left(xAe^{6x}\right) &= 6e^{6x} \\ 12Ae^{6x}(1+3x) - 6Ae^{6x}(1+6x) &= 6e^{6x} \\ 6Ae^{6x} &= 6e^{6x} \\ A &= 1 \end{aligned}$$

Hence,

$$y_p(x) = xe^{6x}$$

Finally, the general solution of the (inhomogeneous) differential equation (3.3.1) is given by

$$\begin{aligned} y(x) &= y_c(x) + y_p(x) \\ &= c_1 + c_2e^{6x} + xe^{6x} \end{aligned}$$

Example 3.3.4 Find the general solution of

$$y'' + y = x + \sin x \tag{3.3.4}$$

Solution. Since the equation is linear (of the form (1.4)) we can write the general solution in the form (see (3.1.1))

$$y(x) = y_c(x) + y_p(x)$$

Complementary Solution y_c

The characteristic equation (see (3.2.2)) is given by

$$\begin{aligned} m^2+1 &= 0 \\ m &= \pm i \end{aligned}$$

From Table 3.2.1 with $a=0$ and $b=1$,

$$y_c(x) = c_1 \sin x + c_2 \cos x \tag{3.3.5}$$

Particular Solution y_p

From (3.3.4), $f(x) = x + \sin x$. Hence, from Table 3.3.1,

$$\begin{aligned} y_p(x) &= \underbrace{y_{p_1}(x)}_{\text{For } x} + \underbrace{y_{p_2}(x)}_{\text{For } \sin x} \\ &= x^{s_1} \underbrace{(B_0 + B_1 x)}_{P_n(x) \text{ with } n=1} + \underbrace{x^{s_2}(A_1 \sin x + A_2 \cos x)}_{\text{For } y_{p_2}(x)} \end{aligned}$$

To determine the required value of s_1, start with the lowest possible value of s_1: $s_1 = 0$ and see if the resulting y_{p_1}, that is, $y_{p_1}(x) = x^0 (B_0 + B_1 x) = B_0 + B_1 x$ solves the corresponding homogeneous differential equation. In this case, we see that no choice of the *constants* c_1 and c_2 in (3.3.5) will lead to a solution of the form $B_0 + B_1 x$. Consequently, $y_{p_1}(x) = B_0 + B_1 x$ is the correct form for $y_{p_1}(x)$.

Repeating this argument for $y_{p_2}(x) = x^{s_2}(A_1 \sin x + A_2 \cos x)$, we find that $s_2 = 1$ and

$$y_{p_2}(x) = x\,(A_1 \sin x + A_2 \cos x)$$

Consequently,

$$\begin{aligned} y_p(x) &= y_{p_1}(x) + y_{p_2}(x) \\ &= B_0 + B_1 x + x\,(A_1 \sin x + A_2 \cos x) \end{aligned} \tag{3.3.6}$$

To find the constants B_0, B_1, A_1 and A_2, we substitute this y_p (from (3.3.6)) back into the inhomogeneous differential equation (3.3.4)

$$\begin{aligned} y_p'' + y_p &= x + \sin x \\ \frac{d^2}{dx^2}\left(B_0 + B_1 x + x\,(A_1 \sin x + A_2 \cos x)\,\right) + B_0 + B_1 x + x\,(A_1 \sin x + A_2 \cos x) &= x + \sin x \\ 2A_1 \cos x - 2A_2 \sin x + B_0 + B_1 x &= x + \sin x \end{aligned}$$

Hence,

$$\begin{aligned} A_1 &= 0,\; A_2 = -\frac{1}{2},\; B_0 = 0,\; B_1 = 1 \\ y_p(x) &= x - \frac{x}{2} \cos x \end{aligned}$$

Finally, the general solution of the (inhomogeneous) differential equation (3.3.4) is given by

$$\begin{aligned} y(x) &= y_c(x) + y_p(x) \\ &= c_1 \sin x + c_2 \cos x + x(1 - \frac{\cos x}{2}) \end{aligned}$$

3.4 Reduction of Order

In Sections 3.2 and 3.3, we discussed the solution of (1.4) in the particular case when the coefficients are constant. Unfortunately, there is no similar procedure for finding routinely the general solution of a higher order differential equation with *variable* coefficients.

Sometimes, however, if we know one solution of a second order homogeneous linear equation with *variable coefficients* (i.e. of the form (3.2.1) with $n = 2$), we can use this information to find the "other linearly independent solution" (see Section 3.1) and hence the general solution (see (3.1.3) with $n = 2$). This is accomplished by reducing the order of the equation by one and solving the resulting first order equation using any one of the methods discussed in Chapter 2. This method is known as *Reduction of Order.* We illustrate by example.

Example 3.4.1 Suppose we know that $y_1 = e^x$ is a solution of

$$xy'' - (x+1)y' + y = 0, \qquad x \neq 0 \tag{3.4.1}$$

We represent the general solution of (3.4.1) in the form

$$y = vy_1 = ve^x \tag{3.4.2}$$

where $v(x)$ is an unknown function to be determined. Substitute (3.4.2) into (3.4.1) first noting that, from (3.4.2),

$$y' = v'e^x + ve^x, \qquad y'' = v''e^x + 2v'e^x + ve^x$$

From (3.4.1) we thus obtain

$$\begin{aligned} x\left(v''e^x + 2v'e^x + ve^x\right) - (x+1)\left(v'e^x + ve^x\right) + ve^x &= 0 \\ v'' + \left(1 - \frac{1}{x}\right)v' &= 0 \end{aligned}$$

Let $w = v'$

$$w' + \left(1 - \frac{1}{x}\right)w = 0 \tag{3.4.3}$$

The resulting differential equation for w(i.e. (3.4.3)) is of first order (hence the order of the original equation (3.4.1) has been reduced by one) and linear in w. Proceeding as in Section 2.5, we obtain (using the integrating factor $\frac{e^x}{x}$)

$$w(x) = cxe^{-x}$$

Since $w = v'$,using integration by parts (Appendix),

$$\begin{aligned} v(x) &= \int cxe^{-x}dx \\ &= c\left[-xe^{-x} - e^{-x}\right] + c_1 \end{aligned}$$

Finally, from (3.4.2), the general solution of (3.4.1) is given by

$$y(x) = ve^x = c_1e^x - c(x+1)$$

Reduction of order can also be used to find particular solutions of *inhomogeneous* differential equations when the method of undetermined coefficients (see Section 3.3) does not apply, that is, when the right-hand side $f(x)$ of the differential equation does not appear in Table 3.3.1. Again, the method relies on a knowledge of *one* solution of the corresponding homogeneous equation. To see this, consider the following example.

Example 3.4.2 Consider, this time, the inhomogeneous differential equation from (3.4.1)

$$xy'' - (x+1)y' + y = x^2(2x+1)e^{(x^2+x)}, \qquad x \neq 0 \qquad (3.4.4)$$

The right-hand side $f(x) = x^2(2x+1)e^{(x^2+x)}$ of (3.4.4) is clearly not in a form which allows for the method of undetermined coefficients (i.e. $f(x)$ does not appear in Table 3.3.1). However, we can still apply reduction of order since, from Example 3.4.1, we know that $y_1 = e^x$ solves the corresponding *homogeneous* equation (3.4.1). Proceeding as in Example 3.4.1, we again seek the general solution of (3.4.4) in the form $y = ve^x$. In this case, instead of (3.4.3), we obtain

$$w' + \left(1 - \frac{1}{x}\right)x = x(2x+1)e^{x^2}$$

which leads to (as in Example 3.4.1)

$$w(x) = xe^{x^2} + cxe^{-x}$$

and

$$v(x) = \int w(x)dx = \int (xe^{x^2} + cxe^{-x})dx$$

Using the substitution rule (Appendix) for the integral $\int xe^{x^2}dx$ (let $u = x^2$) and integration by parts for the integral $\int cxe^{-x}dx$ (as in Example 3.4.1), we obtain

$$v(x) = \frac{e^{x^2}}{2} - ce^{-x}(x+1) + c_1 \tag{3.4.5}$$

Finally, the general solution of (3.4.4) is given by

$$\begin{aligned} y(x) &= ve^x \\ &= \underbrace{\frac{1}{2}e^{(x^2+x)}}_{y_p(x)}\underbrace{-c(x+1) + c_1e^x}_{y_c(x)} \\ &= y_p(x) + y_c(x) \end{aligned}$$

If we set the arbitrary constants of integration c_1 and c_2 to zero (either here or previously in (3.4.5)) we obtain the required particular solution

$$y_p(x) = \frac{1}{2}e^{(x^2+x)}$$

Notice also how the known solution $y_1(x) = e^x$ is "reproduced" as part of y_c, that is, y_c with $c = 0$ and $c_1 = 1$.

Summary of the Method of Reduction of Order

1. The method applies to homogeneous and inhomogeneous linear *second order* ordinary differential equations with *variable* or constant coefficients and can be used to find $y_c(x)$ and (if the equation is inhomogeneous) $y_p(x)$.

2. The method *requires* a single solution $y_1(x)$ of the (corresponding) *homogeneous* equation.

3. The general solution of the homogeneous (or inhomogeneous) equation is then sought in the form
$$y(x) = v(x)y_1(x)$$
where $v(x)$ is an unknown function.

4. The resulting differential equation is first order linear in $v' = w$.

5. Solve for w and integrate to get v.

6. Let $y = v(x)y_1(x)$ to get the general solution of the homogeneous (or inhomogeneous) equation.

7. If only a particular solution $y_p(x)$ of an inhomogeneous equation is required (as in Example 3.4.2), all arbitrary constants of integration should be set to zero.

Note 3.4.3 Reduction of Order can also be applied to linear n^{th} order homogeneous differential equations $(n = 3, 4, \ldots)$ once a single solution $y_1(x)$ is known. The procedure is exactly the same as that in Examples 3.4.1 and 3.4.2. In the case $n \geq 3$, however, the (reduced) resulting equation is of order $n - 1 \geq 2$ and perhaps not as amenable to solution as the *first order* equation resulting from the case $n = 2$.

3.5 Variation of Parameters

The process of finding a particular solution $y_p(x)$ becomes even simpler if we know *even more* information about the corresponding homogeneous equation. In fact, if we know $y_c(x)$ entirely, we can bypass the first order differential equation obtained through the reduction of order process (see Example 3.4.2) and reduce the original inhomogeneous differential equation instead, to a set of *algebraic equations* and an exercise in calculus. This method is known as *Variation of Parameters.* It is particularly useful for finding particular solutions of inhomogeneous linear ordinary differential equations with *constant coefficients* when the right-hand side $f(x)$ is not in a form suitable for the method of undetermined coefficients (i.e. when $f(x)$ does not appear in Table 3.3.1).

We illustrate by example.

Example 3.5.1 Consider the differential equation

$$y'' + y = \tan x \tag{3.5.1}$$

From Example 3.3.4, the complementary solution $y_c(x)$ is given by

$$y_c(x) = c_1 \sin x + c_2 \cos x \tag{3.5.2}$$

Unfortunately, we cannot use the method of undetermined coefficients (§3.3) to find the particular solution $y_p(x)$ since the right-hand side of (3.5.1) does not appear in Table 3.3.1. Instead, use the method of *variation of parameters* as follows.

We suggest a form of particular solution $y_p(x)$ given by (3.5.2) with the constants c_1 and c_2 replaced by the unknown functions $A(x)$ and $B(x)$, respectively:

$$y_p(x) = A(x) \sin x + B(x) \cos x \tag{3.5.3}$$

The objective is to find the functions $A(x)$ and $B(x)$ so that (3.5.3) is indeed a particular solution of (3.5.1). To do this, substitute (3.5.3) into (3.5.1).

1. First, find the first and second derivatives of (3.5.3):

$$y_p'(x) = A'(x)\sin x + B'(x)\cos x + A(x)\cos x - B(x)\sin x$$

   To simplify the second derivative y_p'', we set

$$A'(x)\sin x + B'(x)\cos x = 0 \tag{3.5.4}$$

   Hence,

$$\begin{aligned} y_p''(x) &= \frac{d}{dx}(A(x)\cos x - B(x)\sin x) \\ &= A'(x)\cos x - B'(x)\sin x - (A(x)\sin x + B(x)\cos x) \end{aligned} \tag{3.5.5}$$

2. Substitute $y_p'(x)$ and $y_p''(x)$ into (3.5.1) to obtain

$$A'(x)\cos x - B'(x)\sin x = \tan x \tag{3.5.6}$$

The equations (3.5.4) and (3.5.6) can now be solved simultaneously (using Cramer's rule or elimination) to obtain $A'(x)$ and $B'(x)$. Alternatively, write (3.5.4) and (3.5.6) as a matrix equation (this has certain advantages - see Note 3.5.2 below):

$$\begin{pmatrix} \sin x & \cos x \\ \cos x & -\sin x \end{pmatrix}\begin{pmatrix} A'(x) \\ B'(x) \end{pmatrix} = \begin{pmatrix} 0 \\ \tan x \end{pmatrix} \tag{3.5.7}$$

Multiplying both sides of (3.5.7) by the inverse of the matrix

$$\begin{pmatrix} \sin x & \cos x \\ \cos x & -\sin x \end{pmatrix}$$

we obtain

$$\begin{aligned} \begin{pmatrix} A'(x) \\ B'(x) \end{pmatrix} &= -\begin{pmatrix} -\sin x & -\cos x \\ -\cos x & \sin x \end{pmatrix}\begin{pmatrix} 0 \\ \tan x \end{pmatrix} \\ &= \begin{pmatrix} \sin x & \cos x \\ \cos x & -\sin x \end{pmatrix}\begin{pmatrix} 0 \\ \tan x \end{pmatrix} \end{aligned}$$

i.e.

$$\begin{aligned} A'(x) &= \cos x \tan x = \sin x \\ B'(x) &= -\sin x \tan x = -\frac{\sin^2 x}{\cos x} \end{aligned}$$

Consequently,

$$\begin{aligned} A(x) &= \int \sin x dx = -\cos x \\ B(x) &= \int \frac{\cos^2 x - 1}{\cos x} dx \\ &= \int (\cos x - \sec x)\, dx \\ &= \sin x - \ln|\sec x + \tan x| \end{aligned}$$

(we set the constants of integration to zero since we seek only a particular solution of (3.5.1) - otherwise, we would regenerate y_c, which is known). From (3.5.3)

$$\begin{aligned} y_p(x) &= A(x)\sin x + B(x)\cos x \\ &= (-\cos x)\sin x + (\sin x - \ln|\sec x + \tan x|)\cos x \\ &= -\cos x \ln|\sec x + \tan x| \end{aligned}$$

Finally, since we know $y_c(x)$ (from (3.5.2)), we can write down the general solution of (3.5.1):

$$\begin{aligned} y(x) &= y_c(x) + y_p(x) \\ &= c_1 \sin x + c_2 \cos x - \cos x \ln|\sec x + \tan x| \end{aligned}$$

The essence of the method is summarized as follows.

Summary of the Method of Variation of Parameters

(i) The method can be used to find a particular solution $y_p(x)$ for an inhomogeneous linear *second order* ordinary differential equations with *variable* or constant coefficients:

$$a_2(x)y''(x) + a_1(x)y'(x) + a_0(x)y(x) = f(x), \qquad a_2(x) \neq 0 \tag{3.5.8}$$

(ii) The method *requires* the general solution

$$y_c(x) = c_1 y_1(x) + c_2 y_2(x)$$

of the (corresponding) *homogeneous* equation from (3.5.8) (i.e. $f(x) \equiv 0$ in (3.5.8)). Here, $y_1(x)$ and $y_2(x)$ are the two *known* linearly independent solutions of the corresponding homogeneous equation from (3.5.8) and c_1 and c_2 are the usual arbitrary constants.

(iii) The particular solution of the inhomogeneous equation (3.5.8) is then sought in the form

$$y_p(x) = A(x)y_1(x) + B(x)y_2(x)$$

where $A(x)$ and $B(x)$ are unknown functions (to be determined).

(iv) The resulting system of algebraic equations for $A'(x)$ and $B'(x)$ always takes the form

$$\begin{aligned} A'(x)y_1(x) + B'(x)y_2(x) &= 0 \\ A'(x)y_1'(x) + B'(x)y_2'(x) &= R(x) \end{aligned} \tag{3.5.9a}$$

or, in matrix form,

$$\begin{pmatrix} y_1 & y_2 \\ y_1' & y_2' \end{pmatrix} \begin{pmatrix} A'(x) \\ B'(x) \end{pmatrix} = \begin{pmatrix} 0 \\ R(x) \end{pmatrix} \tag{3.5.9b}$$

where $R(x) = \dfrac{f(x)}{a_2(x)}$.

(v) (3.5.9a) can be solved using either Cramer's rule or elimination. Alternatively, solve the matrix equation (3.5.9b) by first finding the inverse of the matrix $\begin{pmatrix} y_1 & y_2 \\ y_1' & y_2' \end{pmatrix}$ using the formula

$$\begin{pmatrix} y_1 & y_2 \\ y_1' & y_2' \end{pmatrix}^{-1} = \frac{1}{\begin{vmatrix} y_1 & y_2 \\ y_1' & y_2' \end{vmatrix}} \begin{pmatrix} y_2' & -y_2 \\ -y_1' & y_1 \end{pmatrix} \tag{3.5.10}$$

(Note that this inverse always exists since, by Note 3.1.1(b)(ii), since y_1 and y_2 are linearly independent, the determinant (Wronskian)

$$\begin{vmatrix} y_1 & y_2 \\ y_1' & y_2' \end{vmatrix}$$

is never zero on the interval of interest.)

Finally, multiply both sides of (3.5.9b) by the inverse (3.5.10) to obtain

$$\begin{pmatrix} A'(x) \\ B'(x) \end{pmatrix} = \frac{1}{\begin{vmatrix} y_1 & y_2 \\ y_1' & y_2' \end{vmatrix}} \begin{pmatrix} y_2' & -y_2 \\ -y_1' & y_1 \end{pmatrix} \begin{pmatrix} 0 \\ R(x) \end{pmatrix}$$

(vi) Integrate $A'(x)$ and $B'(x)$ to obtain $A(x)$ and $B(x)$ (setting constants of integration to zero).

(vii) Finally, use the expressions for $A(x)$ and $B(x)$ to obtain $y_p(x)$ from

$$y_p(x) = A(x)y_1(x) + B(x)y_2(x)$$

Note 3.5.2 This method is easily extended to equations of higher order. For example, consider the equation

$$a_3(x)y'''(x) + a_2(x)y''(x) + a_1(x)y'(x) + a_0(x)y(x) = f(x), \qquad a_3(x) \neq 0$$

with $y_c(x)$ given by

$$y_c(x) = c_1y_1(x) + c_2y_2(x) + c_3y_3(x)$$

By variation of parameters, suggest $y_p(x)$ of the form

$$y_p(x) = A(x)y_1(x) + B(x)y_2(x) + E(x)y_3(x)$$

This leads to the matrix equation

$$\begin{pmatrix} y_1 & y_2 & y_3 \\ y_1' & y_2' & y_3' \\ y_1'' & y_2'' & y_3'' \end{pmatrix} \begin{pmatrix} A'(x) \\ B'(x) \\ E'(x) \end{pmatrix} = \begin{pmatrix} 0 \\ 0 \\ R(x) \end{pmatrix} \tag{3.5.11}$$

where, $R(x) = \dfrac{f(x)}{a_3(x)}$.

This matrix equation is then solved for $A'(x)$, $B'(x)$ and $E'(x)$, from which are obtained $A(x)$, $B(x)$ and $E(x)$ as before.

Extensions to fourth and higher order equations simply result in the corresponding higher order matrix equation from (3.5.11). For example, for the fourth order differential equation

$$a_4(x)y^{(iv)}(x)+a_3(x)y'''(x)+a_2(x)y''(x)+a_1(x)y'(x)+a_0(x)y(x) = f(x), \qquad a_4(x) \neq 0$$

for a particular solution, we suggest

$$y_p(x) = A(x)y_1(x) + B(x)y_2(x) + E(x)y_3(x) + G(x)y_4(x)$$

and obtain the following matrix equation for A', B', E' and G'.

$$\begin{pmatrix} y_1 & y_2 & y_3 & y_4 \\ y_1' & y_2' & y_3' & y_4' \\ y_1'' & y_2'' & y_3'' & y_4'' \\ y_1''' & y_2''' & y_3''' & y_4''' \end{pmatrix} \begin{pmatrix} A'(x) \\ B'(x) \\ E'(x) \\ G'(x) \end{pmatrix} = \begin{pmatrix} 0 \\ 0 \\ 0 \\ R(x) \end{pmatrix}$$

Here y_i , $i = 1, 2, 3, 4$ are the (known) linearly independent solutions from the given y_c and $R(x) = \dfrac{f(x)}{a_4(x)}$.

3.6 Euler-Cauchy Equations

One special type of linear *variable coefficient* differential equation is the Euler-Cauchy equation which takes the form

$$x^n \frac{d^n y}{dx^n} + a_{n-1} x^{n-1} \frac{d^{n-1} y}{dx^{n-1}} + \ldots + a_1 x \frac{dy}{dx} + a_0 y = 0 \tag{3.6.1}$$

where the $a_0, \ldots, a_{n-1}$ are constant.

The distinguishing feature of (3.6.1) is that, in **each term**, the power of x is the same as the order of the derivative, that is, all terms are of the form

$$x^i \frac{d^i y}{dx^i} \qquad i = 0, 1, 2, \ldots, n-1, n \tag{3.6.2}$$

This *pattern* must be present in *each and every term* in the differential equation - *any* deviation from (3.6.2) means that the equation is not Euler-Cauchy. For example, the equation

$$x^3 \frac{d^3 y}{dx^3} + 2x^2 \frac{d^2 y}{dx^2} + 3\frac{dy}{dx} + 8y = 0$$

is not Euler-Cauchy (i.e. of the form (3.6.1)) since the third term on the left-hand side does not fit the pattern (3.6.2).

To solve an Euler-Cauchy equation, we use the substitution $x = e^t$ to reduce (3.6.1) to a linear constant coefficient differential equation which is then solved as in Section 3.2.

Example 3.6.1 Consider the equation

$$x^2 \frac{d^2 y}{dx^2} - 3x\frac{dy}{dx} + 4y = 0, \qquad x > 0$$

This equation is of the form (3.6.1), that is, Euler-Cauchy. Let $x = e^t$ then

$$\frac{dy}{dx} = e^{-t}\frac{dy}{dt}, \qquad \frac{d^2 y}{dx^2} = e^{-2t}\left(\frac{d^2 y}{dt^2} - \frac{dy}{dt}\right) \tag{3.6.3}$$

Substitute (3.6.3) into the differential equation.

$$\begin{aligned} e^{2t}\left(e^{-2t}\left(\frac{d^2 y}{dt^2} - \frac{dy}{dt}\right)\right) - 3e^t\left(e^{-t}\frac{dy}{dt}\right) + 4y &= 0 \\ \frac{d^2 y}{dt^2} - 4\frac{dy}{dt} + 4y &= 0 \end{aligned}$$

This equation is linear, constant coefficient. As in Section 3.2, the characteristic equation is given by

$$\begin{aligned} m^2 - 4m + 4 &= 0 \\ (m-2)^2 &= 0 \end{aligned}$$

Consequently,

$$y(t) = (c_1 + c_2 t)\, e^{2t}$$

Now, since $x = e^t$, we have $t = \ln x$. Hence,

$$y(x) = (c_1 + c_2 \ln x)\, x^2, \qquad x = e^t > 0$$

Note 3.6.2 Inhomogeneous Euler-Cauchy equations, that is, equations of the form

$$x^n \frac{d^n y}{dx^n} + a_{n-1} x^{n-1} \frac{d^{n-1} y}{dx^{n-1}} + \dots + a_1 x \frac{dy}{dx} + a_0 y = f(x)$$

can be solved as follows:

1. Find $y_c(x)$ as in Example 3.6.1.
2. Find $y_p(x)$ using variation of parameters (§3.5).
3. The general solution is then given by $y(x) = y_c(x) + y_p(x)$.

3.7 Special Types of Second Order Equations

The general second order (not necessarily linear) ordinary differential equation takes the form

$$F(x, y, y', y'') = 0 \tag{3.7.1}$$

In this section, we consider two special classes of second order equations from (3.7.1). These classes are particularly attractive in that they lend themselves to solution by "first order methods"(Chapter 2) - even when (3.7.1) is nonlinear.

Dependent Variable Missing

Suppose the dependent variable y does not appear explicitly in (3.7.1). That is, (3.7.1) takes the form

$$g(x, y', y'') = 0 \tag{3.7.2}$$

By introducing the change of variable

$$y' = p \qquad \text{and} \qquad y'' = \frac{dp}{dx}$$

we can reduce the problem of solving (3.7.2) to that of solving two first order equations in succession.

Example 3.7.1 Solve

$$xy'' + y' = (y')^3, \qquad x \neq 0$$

Solution. Since the dependent variable y does not appear explicitly, we let $y' = p$ and $y'' = \dfrac{dp}{dx}$. The differential equation becomes

$$\begin{aligned} x\frac{dp}{dx} + p &= p^3 \\ x\frac{dp}{dx} &= p^3 - p \end{aligned}$$

This equation is separable (§2.1).

$$\begin{aligned} \int \frac{dp}{p^3 - p} &= \int \frac{dx}{x} \\ \int \frac{dp}{p\,(p-1)\,(p+1)} &= \int \frac{dx}{x} \end{aligned}$$

Using the method of partial fractions (see the Appendix), we have

$$\begin{aligned} \int \left[-\frac{1}{p} + \frac{\frac{1}{2}}{p-1} + \frac{\frac{1}{2}}{p+1} \right] dp &= \ln|x| + \ln A^1 \\ \frac{1}{2}\ln\left|p^2 - 1\right| - \ln|p| &= \ln A\,|x| \\ \ln\left|p^2 - 1\right| - 2\ln|p| &= 2\ln A\,|x| \\ \ln\left|p^2 - 1\right| - \ln|p|^2 &= \ln\left(A\,|x|\right)^2 \\ \ln\left|\frac{p^2 - 1}{p^2}\right| &= \ln\left(A\,|x|\right)^2 \\ \left|\frac{p^2 - 1}{p^2}\right| &= \left(A\,|x|\right)^2 \\ \frac{p^2 - 1}{p^2} &= \pm A^2 x^2 \\ p &= \pm\frac{1}{\sqrt{1 - Cx^2}} \end{aligned}$$

[1] Choosing the integration constant in the form $\ln A$ where $A > 0$.

where $C = \pm A^2$. Now, since $p = \dfrac{dy}{dx}$, we have

$$\begin{aligned} y(x) &= \pm \int \frac{dx}{\sqrt{1 - Cx^2}} \\ &= \begin{cases} \pm\frac{1}{\sqrt{C}} \arcsin\left(x\sqrt{C}\right) + c_1, & C > 0, \\ \\ \pm\frac{1}{\sqrt{-C}} \ln\left|x + \sqrt{\frac{1}{-C} + x^2}\right| + c_2, & C < 0 \end{cases} \end{aligned}$$

Independent Variable Missing

Suppose the independent variable x does not appear explicitly in (3.7.1). That is, (3.7.1) takes the form

$$g(y, y', y'') = 0 \tag{3.7.3}$$

By introducing the change of variable

$$y' = p \qquad \text{and} \qquad y'' = \frac{dp}{dx} = \frac{dp}{dy}\frac{dy}{dx} = p\frac{dp}{dy}$$

we can again reduce the problem of solving (3.7.3) to that of solving two first order equations in succession.

Example 3.7.2 Solve

$$yy'' + 2\left(y'\right)^2 = 0, \qquad y \neq 0$$

Solution. Since the independent variable x does not appear explicitly, we let $y' = p$ and $y'' = p\dfrac{dp}{dy}$. The differential equation becomes

$$\begin{aligned} yp\frac{dp}{dy} + 2p^2 &= 0 \\ p &= 0 \qquad \text{or} \qquad \frac{dp}{dy} + 2\frac{p}{y} = 0 \end{aligned}$$

Case 1 If $p = 0$,

$$\begin{aligned} p &= \frac{dy}{dx} = 0 \\ y &= c_1 \end{aligned}$$

Case 2 If $p \neq 0$, the differential equation is linear in p (§2.5). The integrating factor is given by

$$\exp\left(\int \frac{2}{y}dy\right) = y^2$$

Consequently, the differential equation becomes

$$\begin{aligned} \frac{d}{dy}\left(py^2\right) &= 0 \\ py^2 &= c_2 \\ p &= \frac{c_2}{y^2} \end{aligned}$$

Thus,

$$\frac{dy}{dx} = \frac{c_2}{y^2}$$

This equation is separable (§2.1).

$$\begin{aligned} \int y^2 dy &= c_2 \int dx \\ \frac{y^3}{3} &= c_2 x + c_3 \\ y^3 &= 3c_2 x + 3c_3 \\ &= c_4 x + c_5 \end{aligned}$$

3.8 Series Solutions of Differential Equations

The *series method* can be used to solve a wide class of linear differential equations with *variable coefficients.* Just as the name suggests, solutions are sought in the form of infinite series - which can then be used to compute values of solutions at specific points of interest.

Before we describe the method, we review some basic terminology.

Power Series A power series *in* $(x - x_0)$ (centered at $x = x_0$) is an infinite series of the form

$$\sum_{n=0}^{\infty} b_n (x - x_0)^n \tag{3.8.1}$$

where the coefficients b_n are constant. For the power series (3.8.1), there are only 3 possibilities

(i) The series converges only when $x = x_0$.

(ii) The series converges for all x.

(iii) There is a positive number R such that the series converges for $|x - x_0| < R$ and diverges if $|x - x_0| > R$.

The number R is referred to as the radius of convergence of the series. We adopt the convention that in case (i), $R = 0$ while in case (ii), '$R = \infty$'. For example

$$\frac{1}{1-x} = \sum_{n=0}^{\infty} x^n$$

is a (geometric) series of the form (3.8.1) with $x_0 = 0$ and converges for $|x| < 1$, that is, $R = 1$. On the other hand

$$e^x = \sum \frac{x^n}{n!}$$

is the exponential series and converges for all x ($R = \infty$).

Analytic Function A function $f(x)$ is said to be *analytic* at $x = x_0$ if its power series expansion about x_0:

$$f(x) = \sum_{n=0}^{\infty} b_n (x - x_0)^n \tag{3.8.2}$$

converges to $f(x)$ in some neighborhood of x_0. In this case, the b_n are necessarily of the form

$$b_n = \frac{f^{(n)}(x_0)}{n!}$$

and (3.8.2) is known as the *Taylor series* of $f(x)$ at x_0.

Ordinary Point Consider the differential equation

$$a_2(x)y''(x) + a_1(x)y'(x) + a_0(x)y(x) = 0 \tag{3.8.3}$$

A point x_0 is called an *ordinary point* of (3.8.3) if **both**

$$P(x) = \frac{a_1(x)}{a_2(x)} \quad \text{and} \quad Q(x) = \frac{a_0(x)}{a_2(x)} \tag{3.8.4}$$

are analytic at x_0. If either $P(x)$ and $Q(x)$ is not analytic at x_0, then x_0 is called a *singular point.*

Example 3.8.1 Is $x = 0$ an ordinary or singular point of the following differential equation ?

$$(x-1)\,y'' + x^2 y' + y = 0$$

Solution. Forming

$$P(x) = \frac{x^2}{(x-1)} \quad \text{and} \quad Q(x) = \frac{1}{(x-1)}$$

we note that each of $P(x)$ and $Q(x)$ has its own (Maclaurin's) series which converges near $x = 0$. Hence, $x = 0$ is an ordinary point. Note, however, that $x = 1$ is not an ordinary point since, at $x = 1$, the denominators of $P(x)$ and $Q(x)$ are zero so that neither $P(x)$ nor $Q(x)$ is analytic at $x = 1$. Hence, $x = 1$ is a singular point of the differential equation.

Regular Singular Point The point x_0 is a *regular singular point* of (3.8.3) if x_0 is a singular point and the products

$$(x - x_0)\, P(x) \quad \text{and} \quad (x - x_0)^2\, Q(x)$$

(where P and Q are given by (3.8.4)) are *both analytic* at $x = x_0$. If x_0 is a singular point which is not regular, it is called an *irregular singular point.*

Example 3.8.2 Determine if $x = 0$ and $x = 1$ are ordinary or singular points of the equation

$$\left(x^2 - 1\right)^2 y'' + (x - 1)\, y' + y = 0$$

Solution. Forming

$$P(x) = \frac{(x-1)}{(x-1)^2\,(x+1)^2} \quad \text{and} \quad Q(x) = \frac{1}{(x-1)^2\,(x+1)^2}$$

we note that each of $P(x)$ and $Q(x)$ has its own (Maclaurin's) series which converges near $x = 0$. Hence, $x = 0$ is an ordinary point. Note, however, that $x = 1$ is not an ordinary point since, at $x = 1$, the denominators of $P(x)$ and $Q(x)$ are zero so that neither $P(x)$ nor $Q(x)$ is analytic at $x = 1$. Hence, $x = 1$ is a singular point of the differential equation. However, forming the products

$$\begin{aligned} (x-1)\, P(x) &= \frac{1}{(x+1)^2} \\ (x-1)^2\, Q(x) &= \frac{(x-1)^2}{(x-1)^2\,(x+1)^2} = \frac{1}{(x+1)^2} \end{aligned}$$

we note that $(x-1)\, P(x)$ and $(x-1)^2\, Q(x)$ are both analytic at $x = 1$ so that $x = 1$ is a regular singular point.

Solutions Near an Ordinary Point

We have the following result concerning solutions of a differential equation of the form (from (3.8.3) and (3.8.4))

$$y'' + P(x)y' + Q(x)y = 0 \tag{3.8.5}$$

near an ordinary point.

Theorem 3.8.3 Let x_0 be an ordinary point of the equation (3.8.5). Then (3.8.5) has a solution

$$y(x) = \sum_{n=0}^{\infty} b_n (x - x_0)^n \tag{3.8.6}$$

which contains two arbitrary constants (b_0 and b_1) and which converges (at least) inside a circle with center at $x = x_0$ extending out to the (real or complex) singular point(s) of (3.8.5) nearest $x = x_0$. If (3.8.5) has no singular points in the finite plane, then the solution (3.8.6) is valid for all finite x. The coefficients b_n of (3.8.6) are obtained by substituting (3.8.6) into (3.8.5).

Example 3.8.4 Solve the equation

$$\left(5x^2 - 2\right) y'' + 15xy' + 5y = 0 \tag{3.8.7}$$

near the point $x = 0$.

Solution. The only singular points of (3.8.7) are $x = \pm\sqrt{\dfrac{2}{5}}$. Consequently, $x = 0$ is an ordinary point and, by Theorem 3.8.3, we can represent the general solution of (3.8.7) in the form (3.8.6) with $x_0 = 0$:

$$y(x) = \sum_{n=0}^{\infty} b_n x^n \,, \quad b_0,\ b_1 \text{ arbitrary constants} \tag{3.8.8}$$

with radius of convergence at least $\sqrt{\dfrac{2}{5}}$, that is, the solution from (3.8.8) will be valid for *at least* $|x| < \sqrt{\dfrac{2}{5}}$. To find the b_n's, we substitute (3.8.8) into (3.8.7). To this end, let

$$\begin{aligned} y(x) &= \sum_{n=0}^{\infty} b_n x^n && (3.8.9) \\ y'(x) &= \sum_{n=1}^{\infty} n b_n x^{n-1} && (n = 0 \text{ gives zero contribution}) \\ y''(x) &= \sum_{n=2}^{\infty} n(n-1) b_n x^{n-2} && (n = 0, 1 \text{ give zero contribution}) \end{aligned}$$

Substitute these relations into (3.8.7).

$$\begin{aligned}
(5x^2-2)\sum_{n=2}^{\infty} n(n-1)b_n x^{n-2} + 15x\sum_{n=1}^{\infty} nb_n x^{n-1} + 5\sum_{n=0}^{\infty} b_n x^n &= 0 \\
5\sum_{n=2}^{\infty} n(n-1)b_n x^n - 2\sum_{n=2}^{\infty} n(n-1)b_n x^{n-2} + 15\sum_{n=1}^{\infty} nb_n x^n + 5\sum_{n=0}^{\infty} b_n x^n &= 0
\end{aligned}$$

Noting that (shift of summation)

$$\sum_{n=2}^{\infty} n(n-1)b_n x^{n-2} = \sum_{n=0}^{\infty} (n+1)\,(n+2)b_{n+2}x^n$$

and that the sums in (3.8.9) can be started from $n=0$ without any additional contribution, we obtain

$$\begin{aligned}
\sum_{n=0}^{\infty} \left[5n\,(n-1)\,b_n - 2\,(n+1)\,(n+2)\,b_{n+2} + 15nb_n + 5b_n\right] x^n &= 0 \\
\sum_{n=0}^{\infty} \left[5\left(n^2+2n+1\right) b_n - 2\,(n+1)\,(n+2)\,b_{n+2}\right] x^n &= 0 \\
\sum_{n=0}^{\infty} \left[5\,(n+1)^2\, b_n - 2\,(n+1)\,(n+2)\,b_{n+2}\right] x^n &= 0
\end{aligned}$$

Comparing coefficients of x^n on both sides of this equation, we obtain

$$\begin{aligned}
5\,(n+1)^2\, b_n - 2\,(n+1)\,(n+2)\,b_{n+2} &= 0 \\
b_{n+2} &= \frac{5\,(n+1)^2\, b_n}{2\,(n+1)\,(n+2)}, \quad n \geq 0 \\
b_{n+2} &= \frac{5\,(n+1)\, b_n}{2\,(n+2)}, \quad n \geq 0
\end{aligned}$$

The equation

$$b_{n+2} = \frac{5\,(n+1)\, b_n}{2\,(n+2)}, \quad n \geq 0 \tag{3.8.10}$$

is a two-term recurrence relation for the coefficients b_n , $n \geq 0$ (leading to two arbitrary constants - as expected from Theorem 3.8.3.). Let b_0 and b_1 be arbitrary constants. From 3.8.10,

$$\begin{aligned}
b_0 &= b_0 & b_1 &= b_1\\
b_2 &= \tfrac{5b_0}{2\cdot 2} & b_3 &= \tfrac{5\cdot 2b_1}{2\cdot 3}\\
b_4 &= \tfrac{5\cdot 3b_2}{2\cdot 4} = \tfrac{5^2\cdot 3b_0}{2^2\cdot 2\cdot 4} & b_5 &= \tfrac{5\cdot 4b_3}{2\cdot 5} = \tfrac{5^2\cdot 2\cdot 4b_1}{2^2\cdot 3\cdot 5}\\
b_6 &= \tfrac{5\cdot 5b_4}{2\cdot 6} = \tfrac{5^3\cdot 3\cdot 5b_0}{2^3\cdot 2\cdot 4\cdot 6} & b_7 &= \tfrac{5\cdot 6b_5}{2\cdot 7} = \tfrac{5^3\cdot 2\cdot 4\cdot 6b_1}{2^3\cdot 3\cdot 5\cdot 7}\\
&\vdots & &\vdots\\
b_{2n} &= \tfrac{5^n\cdot 3\cdot 5\cdot 7\ldots(2n-1)}{2^n\cdot 2\cdot 4\ldots(2n)}b_0\,, \quad n\geq 1 & b_{2n+1} &= \tfrac{5^n\cdot 2\cdot 4\cdot 6\ldots(2n)}{2^n\cdot 3\cdot 5\cdot\ldots(2n+1)}b_1\,, \quad n\geq 1
\end{aligned}$$

Hence, the general solution of (3.8.7) is given by

$$\begin{aligned}
y(x) &= \sum_{n=0}^{\infty} b_n x^n\\
&= b_0 + b_1 x + \sum_{n=1}^{\infty} b_{2n}x^{2n} + \sum_{n=1}^{\infty} b_{2n+1}x^{2n+1}\\
&= b_0 + b_1 x + \sum_{n=1}^{\infty} \frac{5^n\cdot 3\cdot 5\cdot 7\ldots(2n-1)}{2^n\cdot 2\cdot 4\cdot\ldots(2n)}b_0 x^{2n} + \sum_{n=1}^{\infty}\frac{5^n\cdot 2\cdot 4\cdot 6\ldots(2n)}{2^n\cdot 3\cdot 5\cdot\ldots(2n+1)}b_1 x^{2n+1}\\
&= b_0\left(1+\sum_{n=1}^{\infty}\frac{5^n\cdot 3\cdot 5\cdot 7\ldots(2n-1)}{2^n\cdot 2\cdot 4\cdot\ldots(2n)}x^{2n}\right) + b_1\left(x+\sum_{n=1}^{\infty}\frac{5^n\cdot 2\cdot 4\cdot 6\ldots(2n)}{2^n\cdot 3\cdot 5\cdot\ldots(2n+1)}x^{2n+1}\right)
\end{aligned}$$

valid, at least, for $|x| < \sqrt{\frac{2}{5}}$ (certainly "near $x = 0$" as required).

Note that the arbitrary constants b_0 and b_1 can be evaluated whenever initial or boundary conditions are given e.g. $y(0) = 1$ and $y'(0) = 0$ imply that $b_0 = 1$ and $b_1 = 0$.

Note 3.8.5

(i) Whenever we wish to obtain solutions near (or about) a point other than $x = 0$, we first translate the problem to that point and then proceed as in Example 3.8.4. For example, if we require the solution of some differential equation near the ordinary point $x = x_0$, we write $u = x - x_0$, transform the differential equation (and any boundary conditions) to one in u and solve this problem (in u) near the ordinary point $u = 0$. Once this problem is solved, we return to the $x-$ domain by reversing the transformation, that is, by replacing u by $x - x_0$.

(ii) The method used in Example 3.8.4 applies also to inhomogeneous differential equations whenever the right-hand side can be expanded in powers of x. In this case, we equate coefficients of x^n, as above, but since the right-hand side of the equation is no longer zero, some of the coefficients b_n of (3.8.8) will include specific numerical values independent of the arbitrary constants b_0 and b_1. This part of the general solution will constitute $y_p(x)$, the particular solution.

Solutions Near a Regular Singular Point

In view of Note 3.8.5(i), we restrict our discussion to the regular singular point $x = 0$. To this end, let $x = 0$ be a regular singular point of the differential equation (3.8.5). For convenience, we write (3.8.5) in the form

$$x^2y'' + xp(x)y' + q(x)y = 0 \tag{3.8.11}$$

where

$$p(x) = xP(x) \quad \text{and} \quad q(x) = x^2Q(x)$$

and, since $x = 0$ is a regular singular point, p and q are analytic at $x = 0$.

In this case, a regular power series method (such as that in Example 3.8.4) will not work. Instead, we seek solutions of (3.8.11) in the form of a *Frobenius series*

$$y(x) = \sum_{n=0}^{\infty} c_n x^{n+r} \tag{3.8.12}$$

where the number r satisfies the quadratic (*indicial*) equation

$$r(r-1) + p(0)r + q(0) = 0 \tag{3.8.13}$$

We have the following result concerning the solutions of (3.8.11) near the regular singular point $x = 0$.

Theorem 3.8.6 Let $x = 0$ be a regular singular point of (3.8.11) and let r_1, r_2 ($r_1 \geq r_2$) be the roots of the indicial equation (3.8.13). For $x > 0$, (3.8.11) has at least one Frobenius series solution corresponding to the larger root r_1 i.e.

$$y_1(x) = \sum_{n=0}^{\infty} a_n x^{n+r_1} \quad , \qquad a_0 \neq 0$$

The second linearly independent solution of (3.8.11) is not necessarily a Frobenius series. In fact, we have the following cases.

(i) If $r_1 - r_2$ is neither zero (i.e. $r_1 \neq r_2$) nor a positive integer, then, for $x > 0$, (3.8.11) has a second linearly independent Frobenius series solution corresponding to the smaller root r_2:

$$y_2(x) = \sum_{n=0}^{\infty} b_n x^{n+r_2} \quad , \qquad b_0 \neq 0$$

(ii) If $r_1 = r_2$, a second linearly independent solution of (3.8.11) takes the form

$$y_2(x) = y_1(x)\ln x + \sum_{n=0}^{\infty} b_n x^{n+r_2+1}$$

(iii) If $r_1 - r_2$ is a positive integer, a second linearly independent solution of (3.8.11) takes the form

$$y_2(x) = Cy_1(x)\ln x + \sum_{n=0}^{\infty} b_n x^{n+r_2}$$

where, $b_0 \neq 0$ but the constant C may be either zero or nonzero.

In each of the above cases (i) - (iii), the series are valid (at least) for $0 < x < R$ where $R > 0$ represents the minimum of the radii of convergence of the (Maclaurin's) series corresponding to the (analytic) functions $p(x) = xP(x)$ and $q(x) = x^2 Q(x)$. The constants a_n, b_n and C are again found by substitution into the differential equation (3.8.11).

Note 3.8.7 To obtain a solution for $x < 0$, we need only replace x by $|x|$ in the solutions obtained from Theorem 3.8.6 for $x > 0$.

Example 3.8.8 Use a series method to find the general solution of the differential equation

$$2xy'' + \left(1 - 2x^2\right) y' - 4xy = 0, \qquad x > 0 \tag{3.8.14}$$

near the point $x = 0$.

Solution. First identify any singular points. From (3.8.5),

$$P(x) = \frac{1 - 2x^2}{2x}, \qquad Q(x) = \frac{-4x}{2x}$$

Hence, $x = 0$ is the only singular point. Next,

$$p(x) = xP(x) = \frac{1}{2}\left(1 - 2x^2\right), \qquad q(x) = x^2 Q(x) = -2x^2 \tag{3.8.15}$$

Since both $p(x)$ and $q(x)$ are analytic at $x = 0$, the latter is a regular singular point. Hence, there exists at least one Frobenius series solution of (3.8.14). To see which form (by Theorem 3.8.6) the second linearly independent solution takes, we examine the indicial equation (3.8.13)

$$\begin{aligned} r(r-1)+p(0)r+q(0) &= 0 \\ r(r-1)+\frac{1}{2}r+0 &= 0 \\ r^2-\frac{1}{2}r &= 0 \\ r\left(r-\frac{1}{2}\right) &= 0 \end{aligned}$$

Hence, the roots of the indicial equation ($r_1=\frac{1}{2}$ and $r_2=0$) differ by a (non-zero) non-integer so that Theorem 3.8.6 (i) applies. That is, there exist two linearly independent Frobenius series solutions

$$\begin{aligned} y_1(x) &= \sum_{n=0}^{\infty} a_n x^{n+\frac{1}{2}}, \quad a_0 \neq 0 \\ y_2(x) &= \sum_{n=0}^{\infty} b_n x^{n+0} \\ &= \sum_{n=0}^{\infty} b_n x^n, \quad b_0 \neq 0 \end{aligned}$$

which (since from (3.8.15) the power series expansions of $p(x)$ and $q(x)$ converge for all $x>0$) are valid $\forall x>0$. We find the coefficients a_n and b_n, at the same time, as follows. Let

$$y(x)=\sum_{n=0}^{\infty} c_n x^{n+r}, \quad c_0 \neq 0$$

where r will (eventually) take the values $\frac{1}{2}$ and 0, as above. Then,

$$y'(x)=\sum_{n=0}^{\infty} c_n\,(n+r)\,x^{n+r-1}, \qquad y''(x)=\sum_{n=0}^{\infty} c_n\,(n+r)\,(n+r-1)\,x^{n+r-2}$$

Substituting into the differential equation (3.8.14), we obtain

$$\begin{aligned} 2\sum_{n=0}^{\infty} c_n\,(n+r)\,(n+r-1)\,x^{n+r-1}+\sum_{n=0}^{\infty} c_n[(n+r)\,x^{n+r-1}-2(n+r)x^{n+r+1}-4x^{n+r+1}] &= 0 \\ \sum_{n=0}^{\infty} c_n\,(n+r)\,(2n+2r-1)x^{n+r-1}-2\sum_{n=0}^{\infty} c_n\,(n+r+2)\,x^{n+r+1} &= 0 \end{aligned}$$

Replace n by $n-2$ in the second sum on the left-hand side:

$$\sum_{n=0}^{\infty} c_n\,(n+r)\,(2n+2r-1)x^{n+r-1}-2\sum_{n=2}^{\infty} c_{n-2}\,(n+r)\,x^{n+r-1}=0$$

Now equate coefficients of different powers of x :

$$\begin{aligned} n &= 0: \quad c_0 r(2r-1) = 0 \text{ or } \quad r(2r-1) = 0 \text{ since } c_0 \neq 0 \\ n &= 1: \quad c_1(1+r)(2r+1) = 0 \text{ or } c_1 = 0 \text{ since } r = 0 \text{ or } \frac{1}{2} \\ n &\geq 2: \quad c_n = \frac{2c_{n-2}}{(2n+2r-1)} \text{ since } (n+r) \neq 0, \quad \text{for } r = 0, \frac{1}{2} \text{ and } n \geq 2 \end{aligned}$$

For each choice of r, we have a separate recurrence relation for the coefficients a_n and b_n.

$$\begin{aligned} \text{When } r &= 0, \text{ from above :} \\ y_2(x) &= \sum_{n=0}^{\infty} b_n x^n, \qquad b_0 \neq 0, \; b_1 = 0 \text{ and } b_n = 2\frac{b_{n-2}}{(2n-1)}, \; n \geq 2 \end{aligned}$$

$$\begin{array}{ll} b_0 = b_0 & b_1 = 0 \\ b_2 = \frac{2b_0}{3} & b_3 = 2\frac{b_1}{5} = 0 \\ b_4 = 2\frac{b_2}{7} = \frac{2^2 b_0}{3 \cdot 7} & b_5 = 0 \\ b_6 = 2\frac{b_4}{11} = \frac{2^3 b_0}{3 \cdot 7 \cdot 11} & b_7 = 0 \\ \vdots & \vdots \\ b_{2k} = \frac{2^k b_0}{3 \cdot 7 \cdot 11 \cdot \ldots (4k-1)}, \quad k \geq 1 & b_{2k+1} = 0, \; k \geq 0 \end{array}$$

It follows that

$$\begin{aligned} y_2(x) &= \sum_{n=0}^{\infty} b_n x^n \\ &= \sum_{n=0}^{\infty} b_{2n} x^{2n} + \sum_{n=0}^{\infty} b_{2n+1} x^{2n+1} \\ &= b_0 + \sum_{n=1}^{\infty} \frac{2^n b_0}{3 \cdot 7 \cdot 11 \cdot \ldots (4n-1)} x^{2n} \end{aligned}$$

$$\begin{aligned} \text{When } r &= \frac{1}{2}, \text{ from above :} \\ y_1(x) &= \sum_{n=0}^{\infty} a_n x^{n+\frac{1}{2}}, \qquad a_0 \neq 0, \; a_1 = 0 \text{ and } a_n = \frac{a_{n-2}}{n}, \; n \geq 2 \end{aligned}$$

$$\begin{array}{ll}
a_0 = a_0 & a_1 = 0 \\
a_2 = \frac{a_0}{2} & a_3 = \frac{a_1}{3} = 0 \\
a_4 = \frac{a_2}{4} = \frac{a_0}{2\cdot 4} & a_5 = 0 \\
a_6 = \frac{a_4}{6} = \frac{a_0}{2\cdot 4\cdot 6} & a_8 = 0 \\
\vdots & \vdots \\
a_{2k} = \frac{a_0}{2\cdot 4\cdot 6\cdot \ldots 2k}, \quad k \geq 1 & a_{2k+1} = 0, \ k \geq 0 \\
a_{2k} = \frac{a_0}{2^k k!}, \quad k \geq 1 &
\end{array}$$

It follows that

$$\begin{aligned}
y_1(x) &= \sum_{n=0}^{\infty} a_n x^{n+\frac{1}{2}} \\
&= \sum_{n=0}^{\infty} a_{2n} x^{2n+\frac{1}{2}} + \sum_{n=0}^{\infty} a_{2n+1} x^{2n+\frac{3}{2}} \\
&= a_0 x^{\frac{1}{2}} + \sum_{n=1}^{\infty} \frac{a_0}{2^n n!} x^{2n+\frac{1}{2}}
\end{aligned}$$

Note that, in this case, $y_1(x)$ simplifies to a recognizable form:

$$\begin{aligned}
y_1(x) &= a_0 x^{\frac{1}{2}} + \sum_{n=1}^{\infty} \frac{a_0}{2^n n!} x^{2n+\frac{1}{2}} \\
&= a_0 x^{\frac{1}{2}} \sum_{n=0}^{\infty} \frac{x^{2n}}{2^n n!} \\
&= a_0 x^{\frac{1}{2}} \sum_{n=0}^{\infty} \frac{\left(\frac{x^2}{2}\right)^n}{n!} \\
&= a_0 x^{\frac{1}{2}} \exp\left(\frac{x^2}{2}\right), \quad x > 0
\end{aligned}$$

Finally, for $x > 0$, the general solution of (3.8.14) is given by

$$\begin{aligned}
y(x) &= y_1(x) + y_2(x) \\
&= a_0 x^{\frac{1}{2}} \exp\left(\frac{x^2}{2}\right) + b_0 (1 + \sum_{n=1}^{\infty} \frac{2^n x^{2n}}{3\cdot 7\cdot 11\cdot \ldots (4n-1)})
\end{aligned}$$

where a_0 and b_0 are (non-zero) arbitrary constants.

Note 3.8.9 - **Point at Infinity** Sometimes we are required to obtain information about the solutions of the equation (3.8.5) for very large values of the independent variable x. When this is the case, we say that we are interested in solutions of (3.8.5) near the *point at infinity ('$x = \infty$')*. Fortunately, it is a relatively simple matter to adapt the above procedures for solutions near $x = 0$ to study solutions near the point at infinity. In fact, the change of variable

$$w = \frac{1}{x} \tag{3.8.16}$$

converts a problem in 'large x' to one in 'small w' (since in (3.8.16) 'large x' corresponds to 'small w'). In other words, the substitution (3.8.16) will convert a problem in x centered around infinity to one in w centered around $w = 0$. Consequently, to solve (3.8.5) near the point at infinity ('$x = \infty$'), we use the following procedure:

1. In (3.8.5) let $w = \dfrac{1}{x}$ and, hence,

$$\begin{aligned} y' &= \frac{dy}{dx} \\ &= \frac{dy}{dw}\frac{dw}{dx} \\ &= \frac{dy}{dw}\left(-\frac{1}{x^2}\right) \\ &= -w^2\frac{dy}{dw} \end{aligned} \tag{3.8.17a}$$

$$\begin{aligned} y'' &= \frac{d}{dx}\left(\frac{dy}{dx}\right) \\ &= \frac{d}{dw}\left(\frac{dy}{dx}\right)\frac{dw}{dx} \\ &= \left(-w^2\frac{d^2y}{dw^2} - 2w\frac{dy}{dw}\right)\left(-w^2\right) \end{aligned} \tag{3.8.17b}$$

The resulting '$w-$ problem' is now analyzed near $w = 0$ (as in Theorems 3.8.3 and 3.8.6). We say that the equation (3.8.5) has an ordinary point, a regular singular point or an irregular singular point at '$x = \infty$' (the point at infinity) if the point $w = 0$ has the corresponding classification for the transformed equation.

2. Once the "$w-$problem" is solved near $w = 0$, we once again make the substitution (3.8.16) in the ($w-$) solution to obtain the solution near the ($x-$) point at infinity.

Example 3.8.10 Solve the equation

$$x^4 y'' + x\left(1 + 2x^2\right) y' + 5y = 0 \tag{3.8.18}$$

for large positive x (near the point at infinity).

Solution. Substituting (3.8.16) and (3.8.17a,b) into the differential equation (3.8.18), we obtain

$$\frac{d^2y}{dw^2} - w\frac{dy}{dw} + 5y = 0 \tag{3.8.19}$$

where y is now a function of the new variable w. The '$w-$equation' (3.8.19) is now analyzed near $w = 0$. In fact, $w = 0$ is an ordinary point of (3.8.19) (as is, therefore, the point '$x = \infty$' of (3.8.18)). From Theorem 3.8.3, we seek the general solution of (3.8.19) in the form of a power series

$$y(w) = \sum_{n=0}^{\infty} b_n w^n$$

and proceed as in Example 3.8.4 to obtain the general solution of (3.8.19)

$$y(w) = b_0 \sum_{n=0}^{\infty} \frac{-15w^{2n}}{2^n n!\,(2n-1)\,(2n-1)\,(2n-5)} + b_1\left(w - \frac{2}{3}w^2 + \frac{1}{15}w^5\right), \quad w > 0$$

Finally, let $w = \dfrac{1}{x}$ to obtain the general solution of (3.8.18) for large positive x (near the point $x = \infty$) :

$$y(x) = b_0 \sum_{n=0}^{\infty} \frac{-15x^{-2n}}{2^n n!\,(2n-1)\,(2n-1)\,(2n-5)} + b_1\left(x^{-1} - \frac{2}{3}x^{-2} + \frac{1}{15}x^{-5}\right), \quad x > 0 \tag{3.8.20}$$

Note that the solution (3.8.20) is constructed around the point at infinity but is, in fact, valid for all $x > 0$.

3.9 Laplace Transform Methods

The Laplace transform is particularly effective in the solution of initial value problems involving linear differential equations with *constant* coefficients. One of the biggest advantages in using the Laplace transform method as opposed to, for example, the method of undetermined coefficients, is that the former can more readily accommodate a wider class of forcing function (right-hand side) in the differential equation: in particular, the class of unit, step or *impulse* functions.

Definition

The Laplace transform of a function $y(t)$, is defined by

$$L[y(t)] = \int_0^\infty e^{-st} y(t)dt = f(s) \tag{3.9.1}$$

The inverse Laplace transform is denoted by

$$y(t) = L^{-1}[f(s)] \tag{3.9.2}$$

Notice that the transformed function, $f(s)$, is a function of the new independent variable s.

Basic Idea when Using Laplace Transforms to solve ODEs

The Laplace transform will transform a linear differential equation with constant coefficients into an *algebraic* equation in the transformed function $f(s)$. The resulting algebraic equation is then solved for the transformed function $f(s)$ which is then inverted (using (3.9.2)) to give a solution $y(t)$ of the differential equation. If appropriate initial conditions are given along with the differential equation e.g. $y(0) = 0$, $y'(0) = 3$, these are automatically incorporated into the solution procedure leading directly to a solution of the *entire* initial value problem (differential equation + initial conditions).

Calculating Laplace transforms from the basic definition (3.9.1), however, is extremely tedious and often complicated. For this reason, it is preferable to use a table of Laplace transforms (see, for example, Table 3.9.1 below) which lists the Laplace transforms of the more common functions arising in applications.

The following table (Table 3.9.1) lists the Laplace transforms of a wide range of functions. The table has been constructed in such a way that it serves two purposes: read from column 1 to column 2, it is a table of Laplace transforms; read from column 2 to column 3, it is a table of *inverse* Laplace transforms.

Table 3.9.1 Table of Laplace Transforms

Function $y(t)$	Transform $L[y(t)] = f(s)$	Inverse Transform $y(t) = L^{-1}[f(s)]$
1	s^{-1}	1
t	s^{-2}	t
$t^n, \quad n = 0, 1, 2, ...$	$s^{-(n+1)}n!$	t^n
$t^a, \quad a > -1$	$s^{-(a+1)}\Gamma(a+1)$	t^a
$(\pi t)^{-\frac{1}{2}}$	$s^{-\frac{1}{2}}$	$(\pi t)^{-\frac{1}{2}}$
e^{at}	$(s-a)^{-1}, \quad s > a$	e^{at}
$t^n e^{at}$	$(s-a)^{-(n+1)} n!, \quad s > a$	$t^n e^{at}$
$\cos kt$	$\frac{s}{s^2+k^2}$	$\cos kt$
$\sin kt$	$\frac{k}{s^2+k^2}$	$\sin kt$
$\cosh kt$	$\frac{s}{s^2-k^2}, \quad s > \lvert k \rvert$	$\cosh kt$
$\sinh kt$	$\frac{k}{s^2-k^2}, \quad s > \lvert k \rvert$	$\sinh kt$
$\sin kt - kt \cos kt$	$\frac{2k^3}{(s^2+k^2)^2}$	$\sin kt - kt \cos kt$
$t \sin kt$	$\frac{2ks}{(s^2+k^2)^2}$	$t \sin kt$
$\sin kt + kt \cos kt$	$\frac{2ks^2}{(s^2+k^2)^2}$	$\sin kt + kt \cos kt$
$\frac{1}{t}(1 - e^{-t})$	$\ln\left(1 + \frac{1}{s}\right)$	$\frac{1}{t}(1 - e^{-t})$
$\frac{2}{t}(1 - \cosh kt)$	$\ln\left(1 - \frac{k^2}{s^2}\right), \quad k^2 < s^2$	$\frac{2}{t}(1 - \cosh kt)$
$\frac{\sin kt}{t}$	$\arctan\left(\frac{k}{s}\right)$	$\frac{\sin kt}{t}$
$\frac{2}{t}(1 - \cos kt)$	$\ln\left(1 + \frac{k^2}{s^2}\right)$	$\frac{2}{t}(1 - \cos kt)$

In Table 3.9.1, a and k are suitable constants, $s > 0$ (unless otherwise stated) and $\Gamma(\cdot)$ denotes the gamma function. The table is organized so that the first two columns constitute a table of *Laplace Transforms* while the last two columns constitute a table of *inverse Laplace transforms.*

In the following table, we summarize the main properties of the Laplace Transform.

Table 3.9.2 Properties of the Laplace Transform

Function $y(t)$	Transform $L[y(t)] = f(s)$	Inverse Transform $L^{-1}[f(s)] = y(t)$
$ay_1(t) + by_2(t)$	$af_1(s) + bf_2(s)$	$ay_1(t) + by_2(t)$
$y'(t)$	$sf(s) - y(0)$	$y'(t)$
$y''(t)$	$s^2 f(s) - sy(0) - y'(0)$	$y''(t)$
$y^{(n)}(t)$	$s^n f(s) - s^{n-1}y(0) - \cdots - y^{(n-1)}(0)$	$y^{(n)}(t)$
$\int_0^t y(\beta)d\beta$	$\frac{f(s)}{s}$	$\int_0^t y(\beta)d\beta$
$e^{at}y(t)$	$f(s-a)$	$e^{at}y(t)$
$\alpha(t-c) = \begin{cases} 0, & 0 \le t < c, \\ 1, & t \ge c \end{cases}$	$s^{-1}e^{-cs}, \quad c > 0$	$\alpha(t-c) = \begin{cases} 0, & 0 \le t < c \\ 1, & t \ge c \end{cases}$
$y(t-c)\,\alpha(t-c)$	$e^{-cs}f(s), \quad c > 0$	$y(t-c)\,\alpha(t-c)$
$\int_0^t y_1(\beta)y_2(t-\beta)\,d\beta$	$f_1(s)\,f_2(s)$	$\int_0^t y_1(\beta)y_2(t-\beta)\,d\beta$
$ty(t)$	$-f'(s)$	$ty(t)$
$t^n y(t)$	$(-1)^n f^{(n)}(s)$	$t^n y(t)$
$\frac{y(t)}{t}$	$\int_s^\infty f(\beta)d\beta$	$\frac{y(t)}{t}$
$y(t)$, period p	$\frac{1}{1-e^{-ps}} \int_0^p e^{-st}y(t)dt$	$y(t)$, period p

Here, a,b and c are suitably chosen constants and $s > 0$.

Unfortunately, the majority of Laplace transforms cannot be inverted by inspection, that is, immediately from tables. In most cases we must first reduce the given transform to a *recognizable* or *standard* form and then use tables of Laplace transforms. The situation is similar to that arising in the theory of integration (or anti-differentiation) where a variety of techniques such as *integration by substitution, partial fractions, completing the square* etc. are used to reduce a "complicated integral" to a standard one appearing in a table of integrals. In fact, most of these same techniques can be used when inverting Laplace transforms.

Example 3.9.3 Suppose we are required to find $L^{-1}\left[\dfrac{1}{s^2+2s+5}\right]$. We note that no function of this form appears in either of Tables 3.9.1 or 3.9.2. However, completing the square (see the Appendix) in the denominator yields

$$s^2+2s+5=(s+1)^2+4$$

Hence,

$$\frac{1}{s^2+2s+5}=\frac{1}{(s+1)^2+4}=f\,(s+1)$$

where

$$f\,(\cdot)=\frac{1}{(\cdot)^2+4}$$

Comparing this with entry 7 ($a=-1$) in Table 3.9.2, we find that the inverse transform is given by

$$\begin{aligned} L^{-1}\left[\frac{1}{s^2+2s+5}\right] &= L^{-1}\left[\frac{1}{(s+1)^2+4}\right] \\ &= L^{-1}\left[f\,(s+1)\right] \\ &= e^{-t}L^{-1}\left[f(s)\right] \\ &= e^{-t}L^{-1}\left[\frac{1}{s^2+4}\right] \\ &= e^{-t}\frac{1}{2}\sin 2t \quad \text{(Table 3.9.1)} \end{aligned}$$

Many other examples will be found in the examinations in Part 2 of the text.

Consider next, the solution of an initial value problem using the Laplace transform method.

Example 3.9.4 Solve the initial value problem

$$y''(t)+3y'(t)+2y(t)=4t^2, \qquad y(0)=y'\,(0)=0$$

using Laplace transforms.

Solution. Apply the Laplace transform to both sides of the differential equation.

$$L[y''(t)] + 3L[y'(t)] + 2L[y(t)] = 4L[t^2]$$

From Table 3.9.2 we have

$$s^2 L[y(t)] - sy(0) - y'(0) + 3(sL[y(t)] - y(0)) + 2L[y(t)] = 4.\frac{2}{s^3}$$

Applying the given initial conditions,

$$\begin{aligned} (s^2 + 3s + 2)L[y(t)] &= \frac{8}{s^3} \\ L[y(t)] &= \frac{8}{s^3(s+2)(s+1)} \end{aligned}$$

Next, invert the transform. Before we can do this, it is necessary to decompose the expression $\frac{8}{s^3(s+2)(s+1)}$ into (standard) transforms which can be found from Table 3.9.1. We use the method of partial fractions (see Appendix).

$$\frac{8}{s^3(s+2)(s+1)} = \frac{4}{s^3} - \frac{6}{s^2} + \frac{7}{s} + \frac{1}{s+2} - \frac{8}{s+1}$$

Hence,

$$\begin{aligned} y(t) &= L^{-1}\left[\frac{8}{s^3(s+2)(s+1)}\right] \\ &= L^{-1}\left[\frac{4}{s^3} - \frac{6}{s^2} + \frac{7}{s} + \frac{1}{s+2} - \frac{8}{s+1}\right] \\ &= 4L^{-1}[\frac{1}{s^3}] - 6L^{-1}[\frac{1}{s^2}] + 7L^{-1}[\frac{1}{s}] + L^{-1}[\frac{1}{s+2}] - 8L^{-1}[\frac{1}{s+1}] \end{aligned}$$

From Table 3.9.1, we have

$$\begin{aligned} y(t) &= 4\left(\frac{t^2}{2!}\right) - 6\left(\frac{t}{1!}\right) + 7 + e^{-2t} - 8e^{-t} \\ &= 2t^2 - 6t + 7 + e^{-2t} - 8e^{-t} \end{aligned}$$

The method of Laplace Transforms can also be used to solve systems of equations.

Example 3.9.5 Solve the system

$$\begin{aligned} 2y''(t) + x(t) + 2y(t) &= 0 \\ x'(t) + 2y'(t) &= 0 \\ y(0) &= y'(0) = 0, \quad x(0) = 3 \end{aligned}$$

Solution. Take the Laplace transform of both differential equations using Table 3.9.1 and the given initial conditions.

$$\begin{aligned} 2s^2 L[y(t)] + L[x(t)] + 2L[y(t)] &= 0 \\ sL[x(t)] - 3 + 2sL[y(t)] &= 0 \end{aligned}$$

i.e.

$$\begin{aligned} 2\left(s^2+1\right) L[y(t)] + L[x(t)] &= 0 \\ L[x(t)] + 2L[y(t)] &= \frac{3}{s} \end{aligned}$$

Solve this last system to find:

$$L[y(t)] = -\frac{3}{2s^3}, \qquad L[x(t)] = \frac{3}{s} + \frac{3}{s^3}$$

Finally, invert both transforms to obtain

$$y(t) = -\frac{3}{4}t^2, \qquad x(t) = 3\left(1 + \frac{1}{2}t^2\right)$$

3.10 Linear Systems of Equations: Eigenvalue Methods

There are many different methods for solving linear systems of differential equations with constant coefficients. For example, *elimination methods, matrix techniques* or *Laplace transform methods* (see Example 3.9.5). A full account of each of these methods can be found in most textbooks dealing with the theory of differential equations.

In this section, we turn our attention to one of the most powerful techniques for solving *first order* linear systems of differential equations with constant coefficients - the *eigenvalue method.*

THE EIGENVALUE METHOD FOR HOMOGENEOUS SYSTEMS

Consider the following homogeneous first order linear system with constant coefficients a_{ij} , $i, j = 1, ...n$.

$$\begin{aligned} x_1'(t) &= a_{11}x_1(t) + a_{12}x_2(t) + \cdots + a_{1n}x_n(t) \\ x_2'(t) &= a_{21}x_1(t) + a_{22}x_2(t) + \cdots + a_{2n}x_n(t) \\ &\vdots \\ x_n'(t) &= a_{n1}x_1(t) + a_{n2}x_2(t) + \cdots + a_{nn}x_n(t) \end{aligned} \tag{3.10.1}$$

We rewrite this system in matrix form

$$\mathbf{x}'(t) = A\mathbf{x}(t) \tag{3.10.2}$$

where $\mathbf{x}(t)$ is the vector (or $(n \times 1) - matrix$) given by

$$\begin{bmatrix} x_1(t) \\ x_2(t) \\ \vdots \\ x_n(t) \end{bmatrix}$$

A is the $(n \times n)-$ (constant) coefficient matrix $[a_{ij}]$ and $\mathbf{x}'(t) = \dfrac{d\mathbf{x}(t)}{dt}$.
Closely related to the general solution of (3.10.2) is the concept of *eigenvalue* and *eigenvector.*

Definition

The number λ is called an *eigenvalue* or characteristic value of the matrix A whenever

$$\det(A - \lambda I) = 0 \tag{3.10.3}$$

The associated *eigenvector* is the nonzero vector (or $(n \times 1) - matrix$) $\mathbf{v}$ such that

$$(A - \lambda I)\,\mathbf{v} = \mathbf{0} \tag{3.10.4}$$

In (3.10.3) and (3.10.4), I is the $(n \times n)-$ identity matrix defined by

$$I = \begin{bmatrix} 1 & 0 & 0 & 0 & \cdots & 0 \\ 0 & 1 & 0 & 0 & \cdots & 0 \\ 0 & 0 & 1 & 0 & \cdots & 0 \\ 0 & 0 & 0 & 1 & \cdots & 0 \\ \vdots & \vdots & \vdots & \vdots & \ddots & \vdots \\ 0 & 0 & 0 & 0 & \cdots & 1 \end{bmatrix}$$

Regarding solutions of (3.10.1), we have the following result.

Theorem 3.10.1

(i) The general solution of the system (3.10.1) takes the form

$$\mathbf{x}(t) = c_1\mathbf{x}_1(t) + c_2\mathbf{x}_2(t) + \cdots + c_n\mathbf{x}_n(t) \tag{3.10.5}$$

where $\mathbf{x}_i(t)$, $i = 1, ...n$ are n *linearly independent* solution vectors/matrices of the system (3.10.1) and c_i, $i = 1, ...n$ are arbitrary constants.

(ii) If λ is an eigenvalue of A with associated eigenvector $\mathbf{v}$, then

$$\mathbf{x}(t) = \mathbf{v}e^{\lambda t} \tag{3.10.6}$$

is a nontrivial solution of (3.10.1).

Clearly, from Theorem 3.10.1, if we can find n linearly independent eigenvectors $\mathbf{v}_i$, $i = 1, ...n$, we can form n linearly independent solutions of the form (3.10.6) and use (3.10.5) to find the general solution of the system (3.10.1).

Hence, the general solution of (3.10.1) depends on the eigenvalues and eigenvectors of the matrix A and hence on solutions of the equation (3.10.2) - much in the same way that the general solution of the linear homogeneous ordinary differential equation (3.2.1) depended on the solutions of the associated characteristic equation (3.2.2). For this reason, the equation (3.10.2) is often referred to as the *characteristic equation* for the system (3.10.1).

In fact, the availability of the required n linearly independent eigenvectors $\mathbf{v}_i$, $i = 1, ...n$ in (3.10.5) depends heavily on the nature of the solutions of the equation (3.10.2), that is, on the nature of the *eigenvalues.* As in the case of the characteristic equation (3.2.2), the solutions of (3.10.2) (i.e. the eigenvalues) can be real or complex, distinct or repeated. This introduces difficulties when trying to find the required number of linearly independent eigenvectors $\mathbf{v}_i$, $i = 1, ...n$ for (3.10.5).

We summarize the relevant results for each of the cases when the eigenvalues are distinct or repeated, real or complex.

Distinct Real Eigenvalues

Suppose matrix A has n real and distinct eigenvalues $\lambda_1, \lambda_2, ..., \lambda_n$. Substituting each of these into (3.10.4) and solving for the corresponding eigenvectors $\mathbf{v}_1, \mathbf{v}_2, ..., \mathbf{v}_n$ leads to n linearly independent solutions of the form (3.10.6). The general solution of (3.10.1) is then given by

$$\mathbf{x}(t) = c_1\mathbf{v}_1e^{\lambda_1 t} + c_2\mathbf{v}_2e^{\lambda_2 t} + ... + c_n\mathbf{v}_ne^{\lambda_n t} \tag{3.10.7}$$

where $c_i, i = 1, ...n$, are arbitrary constants.

Example 3.10.2 Find the general solution of the system

$$\begin{aligned} x_1' &= 2x_1 + x_2 \\ x_2' &= x_1 + 2x_2 \end{aligned}$$

Solution. Writing the system in matrix form we obtain

$$\mathbf{x}'(t) = \begin{bmatrix} 2 & 1 \\ 1 & 2 \end{bmatrix} \mathbf{x}(t)$$

Hence, the matrix A is given by

$$A = \begin{bmatrix} 2 & 1 \\ 1 & 2 \end{bmatrix}$$

The characteristic equation (3.10.3) is

$$\begin{aligned} \begin{vmatrix} 2-\lambda & 1 \\ 1 & 2-\lambda \end{vmatrix} &= (2-\lambda)^2 - 1 \\ &= 3 - 4\lambda + \lambda^2 \\ &= (\lambda - 3)(\lambda - 1) \\ &= 0 \end{aligned}$$

Hence, we have real, distinct eigenvalues given by $\lambda_1 = 1$, $\lambda_2 = 3$. The corresponding eigenvector equation (3.10.4) is given by

$$\begin{bmatrix} 2-\lambda & 1 \\ 1 & 2-\lambda \end{bmatrix} \begin{bmatrix} v_1 \\ v_2 \end{bmatrix} = \begin{bmatrix} 0 \\ 0 \end{bmatrix} \tag{3.10.8}$$

Let $\lambda = \lambda_1 = 1$.

The eigenvector equation becomes

$$\begin{bmatrix} 1 & 1 \\ 1 & 1 \end{bmatrix} \begin{bmatrix} v_1 \\ v_2 \end{bmatrix} = \begin{bmatrix} 0 \\ 0 \end{bmatrix}$$

We have the equation

$$v_1 + v_2 = 0 \quad \text{(twice)}$$

We choose $v_1 = c =$ an arbitrary (but nonzero) constant and solve for v_2: $v_2 = -v_1 = -c$. Hence, the eigenvector is given by

$$\begin{bmatrix} v_1 \\ v_2 \end{bmatrix} = c \begin{bmatrix} 1 \\ -1 \end{bmatrix}$$

Let $c = 1$ (for convenience) and we obtain the eigenvector

$$\mathbf{v}_1 = \begin{bmatrix} 1 \\ -1 \end{bmatrix}$$

corresponding to the eigenvector $\lambda_1 = 1$.

Let $\lambda = \lambda_2 = 3$.

The eigenvector equation (3.10.8) becomes

$$\begin{bmatrix} -1 & 1 \\ 1 & -1 \end{bmatrix} \begin{bmatrix} v_1 \\ v_2 \end{bmatrix} = \begin{bmatrix} 0 \\ 0 \end{bmatrix}$$

We have the equations

$$\begin{aligned} -v_1 + v_2 &= 0 \\ v_1 - v_2 &= 0 \end{aligned}$$

These equations are equivalent (notice that one is $-1\times$ the other). We choose v_1 $= c =$ an arbitrary (but nonzero) constant and solve for v_2: $v_2 = v_1 = c$. Hence, the eigenvector is given by

$$\begin{bmatrix} v_1 \\ v_2 \end{bmatrix} = c \begin{bmatrix} 1 \\ 1 \end{bmatrix}$$

Let $c = 1$ (for convenience) and we obtain the eigenvector

$$\mathbf{v}_2 = \begin{bmatrix} 1 \\ 1 \end{bmatrix}$$

corresponding to the eigenvector $\lambda_1 = 3$.

From (3.10.6), these eigenvalues and eigenvectors lead to the two linearly independent solutions

$$\mathbf{x}_1 = \begin{bmatrix} 1 \\ -1 \end{bmatrix} e^t, \qquad \mathbf{x}_2 = \begin{bmatrix} 1 \\ 1 \end{bmatrix} e^{3t}$$

so that the general solution of the system is given by (from (3.10.7))

$$\begin{aligned} \mathbf{x}(t) &= \mathbf{v}_1 e^{\lambda_1 t} + \mathbf{v}_2 e^{\lambda_2 t} \\ &= c_1 \begin{bmatrix} 1 \\ -1 \end{bmatrix} e^t + c_2 \begin{bmatrix} 1 \\ 1 \end{bmatrix} e^{3t} \end{aligned}$$

or

$$x_1(t) = c_1 e^t + c_2 e^{3t}, \qquad x_2(t) = -c_1 e^t + c_2 e^{3t}$$

where c_1 and c_2 are arbitrary constants.

Complex Eigenvalues

Since matrix A is real, the characteristic equation will always have real coefficients. Hence, any complex (non-real) eigenvalues will occur in conjugate pairs. Suppose that $\lambda = a + ib$, $\bar{\lambda} = a - ib$ is such a pair of complex eigenvalues. Denote by $\mathbf{v} = \mathbf{p} + i\mathbf{q}$ and $\bar{\mathbf{v}} = \mathbf{p} - i\mathbf{q}$ the corresponding (complex) eigenvectors. Then, the corresponding *real* solutions of the system (3.10.1) can be shown to take the form

$$\begin{aligned} \mathbf{x}_1(t) &= \mathrm{Re}\left(\mathbf{v}e^{\lambda t}\right) = e^{at}(\mathbf{p}\cos at - \mathbf{q}\sin bt) \\ \mathbf{x}_2(t) &= \mathrm{Im}\left(\mathbf{v}e^{\lambda t}\right) = e^{at}(\mathbf{q}\cos at + \mathbf{p}\sin bt) \end{aligned} \tag{3.10.9}$$

Example 3.10.3 Find the general solution of the system

$$\begin{aligned} x_1' &= 2x_1 + x_2 \\ x_2' &= -4x_1 + 2x_2 \end{aligned}$$

Solution. Writing the system in matrix form we obtain

$$\mathbf{x}'(t) = \begin{bmatrix} 2 & 1 \\ -4 & 2 \end{bmatrix} \mathbf{x}(t)$$

Hence, the matrix A is given by

$$A = \begin{bmatrix} 2 & 1 \\ -4 & 2 \end{bmatrix}$$

The characteristic equation (3.10.3) is

$$\begin{aligned} \begin{vmatrix} 2-\lambda & 1 \\ -4 & 2-\lambda \end{vmatrix} &= (2-\lambda)^2 + 4 \\ &= 8 - 4\lambda + \lambda^2 \\ &= 0 \end{aligned}$$

Hence, we have complex eigenvalues given by $\lambda_1 = 2 + 2i$, $\lambda_2 = \bar{\lambda}_1 = 2 - 2i$. The corresponding eigenvector equation (3.10.4) is given by

$$\begin{bmatrix} 2-\lambda & 1 \\ -4 & 2-\lambda \end{bmatrix} \begin{bmatrix} v_1 \\ v_2 \end{bmatrix} = \begin{bmatrix} 0 \\ 0 \end{bmatrix} \tag{3.10.10}$$

Let $\lambda = \lambda_1 = 2 + 2i$.

The eigenvector equation becomes

$$\begin{bmatrix} -2i & 1 \\ -4 & -2i \end{bmatrix} \begin{bmatrix} v_1 \\ v_2 \end{bmatrix} = \begin{bmatrix} 0 \\ 0 \end{bmatrix}$$

We have the equations

$$\begin{aligned} -2iv_1 + v_2 &= 0 \\ -4v_1 - 2iv_2 &= 0 \end{aligned}$$

These equations are equivalent (i.e. one is $-2i\times$ the other). We choose $v_1 = c =$ an arbitrary (but nonzero) constant and solve for v_2: $v_2 = 2iv_1 = 2ic$. Hence, the eigenvector is given by

$$\begin{bmatrix} v_1 \\ v_2 \end{bmatrix} = c \begin{bmatrix} 1 \\ 2i \end{bmatrix}$$

Let $c = 1$ (for convenience) and we obtain the eigenvector

$$\mathbf{v}_1 = \begin{bmatrix} 1 \\ 2i \end{bmatrix} = \begin{bmatrix} 1 \\ 0 \end{bmatrix} + i \begin{bmatrix} 0 \\ 2 \end{bmatrix}$$

corresponding to the eigenvector $\lambda_1 = 2 + 2i$.

Let $\lambda = \lambda_2 = 2 - 2i$.

From what has been said above, we know that the eigenvector corresponding to this eigenvalue will take the (conjugate) form

$$\bar{\mathbf{v}}_1 = \begin{bmatrix} 1 \\ -2i \end{bmatrix} = \begin{bmatrix} 1 \\ 0 \end{bmatrix} - i \begin{bmatrix} 0 \\ 2 \end{bmatrix}$$

From (3.10.9) with $\mathbf{p} = \begin{bmatrix} 1 \\ 0 \end{bmatrix}$ and $\mathbf{q} = \begin{bmatrix} 0 \\ 2 \end{bmatrix}$, these eigenvalues and eigenvectors lead to the two linearly independent solutions

$$\mathbf{x}_1 = e^{2t}\left(\begin{bmatrix} 1 \\ 0 \end{bmatrix} \cos 2t - \begin{bmatrix} 0 \\ 2 \end{bmatrix} \sin 2t\right), \qquad \mathbf{x}_2 = e^{2t}\left(\begin{bmatrix} 0 \\ 2 \end{bmatrix} \cos 2t + \begin{bmatrix} 1 \\ 0 \end{bmatrix} \sin 2t\right)$$

so that the general solution of the system is given by (from (3.10.5))

$$\begin{aligned} \mathbf{x}(t) &= c_1\mathbf{x}_1(t) + c_2\mathbf{x}_2(t) \\ &= c_1 e^{2t}\left(\begin{bmatrix} 1 \\ 0 \end{bmatrix} \cos 2t - \begin{bmatrix} 0 \\ 2 \end{bmatrix} \sin 2t\right) + c_2 e^{2t}\left(\begin{bmatrix} 0 \\ 2 \end{bmatrix} \cos 2t + \begin{bmatrix} 1 \\ 0 \end{bmatrix} \sin 2t\right) \end{aligned}$$

or

$$x_1(t) = e^{2t}\left(c_1 \cos 2t + c_2 \sin 2t\right), \qquad x_2(t) = e^{2t}\left(-2c_1 \sin 2t + 2c_2 \cos 2t\right)$$

where c_1 and c_2 are arbitrary constants.

Repeated Eigenvalues

We confine our discussion to the case when the repeated eigenvalue has multiplicity $k = 2$ - although what follows is true also for eigenvalues of multiplicity $k > 2$. If an eigenvalue λ has multiplicity 2 (i.e. λ occurs twice as a root of the characteristic equation (3.10.3)), there is the possibility that we obtain less than 2 linearly independent eigenvectors corresponding to λ. This in turn means that the eigenvalue method *may* produce *fewer* than the required $n \geq 2$ linearly independent solutions of the system (3.10.2). There are two cases to consider:

Case 1 An eigenvalue λ has multiplicity 2 but there continues to be two corresponding linearly independent eigenvectors and hence the required two linearly independent solutions of the system. This is always the case when the matrix A in (3.10.2) is symmetric.

Case 2 An eigenvalue λ has multiplicity 2 but there is only one corresponding eigenvector and hence only one corresponding solution of the system - in this case, we have to find another linearly independent solution of the system. We do this in much the same way as we did for a repeated root of the characteristic equation (3.2.2) for the linear homogeneous ordinary differential equation (3.2.1).

We discuss each of these cases separately.

Case 1 Consider the system

$$\begin{aligned} x_1' &= 2x_1 + 3x_2 + 3x_3 \\ x_2' &= \quad -x_2 - 3x_3 \\ x_3' &= \quad 2x_3 \end{aligned}$$

The coefficient matrix A is given by

$$A = \begin{bmatrix} 2 & 3 & 3 \\ 0 & -1 & -3 \\ 0 & 0 & 2 \end{bmatrix}$$

The characteristic equation is given by

$$\begin{aligned} \begin{vmatrix} 2-\lambda & 3 & 3 \\ 0 & -1-\lambda & -3 \\ 0 & 0 & 2-\lambda \end{vmatrix} &= -(1+\lambda)(2-\lambda)^2 = 0 \\ \lambda_1 &= -1, \quad \lambda_2 = 2 \text{ (twice)} \end{aligned}$$

Hence, we have the real eigenvalue $\lambda_1 = -1$ and the repeated eigenvalue $\lambda_2 = 2$ of multiplicity 2. Following the procedure in Examples 3.10.2, 3.10.3 above, we find that $\lambda_1 = -1$ has associated eigenvector $\mathbf{v}_1 = \begin{bmatrix} 1 \\ -1 \\ 0 \end{bmatrix}$. When $\lambda_2 = 2$, the eigenvector equation (3.10.4) becomes

$$\begin{bmatrix} 2-\lambda & 3 & 3 \\ 0 & -1-\lambda & -3 \\ 0 & 0 & 2-\lambda \end{bmatrix} \begin{bmatrix} v_1 \\ v_2 \\ v_3 \end{bmatrix} = \begin{bmatrix} 0 \\ 0 \\ 0 \end{bmatrix}$$

$$\begin{bmatrix} 0 & 3 & 3 \\ 0 & -3 & -3 \\ 0 & 0 & 0 \end{bmatrix} \begin{bmatrix} v_1 \\ v_2 \\ v_3 \end{bmatrix} = \begin{bmatrix} 0 \\ 0 \\ 0 \end{bmatrix}$$

This is equivalent to the single scalar equation

$$\begin{aligned} v_2 + v_3 &= 0 \\ v_1 &= c_1 = \text{arbitrary} \end{aligned}$$

Let $v_2 = c_2 =$ arbitrary. We have

$$\begin{bmatrix} v_1 \\ v_2 \\ v_3 \end{bmatrix} = \begin{bmatrix} c_1 \\ c_2 \\ -c_2 \end{bmatrix} = c_1 \begin{bmatrix} 1 \\ 0 \\ 0 \end{bmatrix} + c_2 \begin{bmatrix} 0 \\ 1 \\ -1 \end{bmatrix}$$

Thus we have the *two* linearly independent eigenvectors

$$\begin{aligned} \mathbf{v}_2 &= \begin{bmatrix} 1 \\ 0 \\ 0 \end{bmatrix} \\ \mathbf{v}_3 &= \begin{bmatrix} 0 \\ 1 \\ -1 \end{bmatrix} \end{aligned}$$

corresponding to $\lambda_2 = 2$. Finally, the three linearly independent eigenvectors $\mathbf{v}_i$, $i = 1, 2, 3$ lead to the general solution of the system:

$$\begin{aligned} \mathbf{x}(t) &= \mathbf{v}_1 e^{\lambda_1 t} + \mathbf{v}_2 e^{\lambda_2 t} + \mathbf{v}_3 e^{\lambda_2 t} \\ &= c_1 \begin{bmatrix} 1 \\ -1 \\ 0 \end{bmatrix} e^{-t} + c_2 \begin{bmatrix} 1 \\ 0 \\ 0 \end{bmatrix} e^{2t} + c_3 \begin{bmatrix} 0 \\ 1 \\ -1 \end{bmatrix} e^{2t} \end{aligned}$$

or

$$x_1(t) = c_1 e^{-t} + c_2 e^{2t}, \qquad x_2(t) = -c_1 e^{-t} + c_3 e^{2t}, \qquad x_3(t) = -c_3 e^{2t}$$

Case 2 Suppose the eigenvalue λ has multiplicity 2. We continue to have the 'first' solution corresponding to λ :

$$\mathbf{x}_1(t) = \mathbf{v}_1 e^{\lambda t}$$

where $\mathbf{v}_1$ is an eigenvector associated with λ_1. The 'missing' solution $\mathbf{x}_2(t)$ is constructed as follows.

$$\mathbf{x}_2(t) = e^{\lambda t} (\mathbf{v}_1 t + \mathbf{v}_2) \tag{3.10.11}$$

where, this time,

$$(A - \lambda I)\, \mathbf{v}_2 = \mathbf{v}_1 \tag{3.10.12}$$

Despite the fact that, since λ is an eigenvalue, $|A - \lambda I| = 0$, it can be shown that (3.10.12) can always be solved for $\mathbf{v}_2$.

Example 3.10.4 Consider the system with coefficient matrix

$$A = \begin{bmatrix} 4 & 1 \\ -1 & 2 \end{bmatrix}$$

The characteristic equation is

$$\begin{aligned} \begin{vmatrix} 4-\lambda & 1 \\ -1 & 2-\lambda \end{vmatrix} &= \\ (4-\lambda)(2-\lambda)+1 &= 0 \\ \lambda^2 - 6\lambda + 9 &= 0 \\ (\lambda-3)^2 &= 0 \end{aligned}$$

Hence, we have the repeated eigenvalue $\lambda = 3$ of multiplicity 2. With this value of λ, we have the following eigenvector equation

$$\begin{bmatrix} 1 & 1 \\ -1 & -1 \end{bmatrix} \begin{bmatrix} v_1 \\ v_2 \end{bmatrix} = \begin{bmatrix} 0 \\ 0 \end{bmatrix}$$

or the single scalar equation $v_1 + v_2 = 0$. As before, this yields the eigenvector and solution

$$\mathbf{v}_1 = \begin{bmatrix} 1 \\ -1 \end{bmatrix}, \qquad \mathbf{x}_1(t) = \begin{bmatrix} 1 \\ -1 \end{bmatrix} e^{3t}$$

Equation (3.10.12) now takes the form

$$\begin{aligned} (A - \lambda I)\mathbf{v}_2 &= \mathbf{v}_1 \\ \begin{bmatrix} 1 & 1 \\ -1 & -1 \end{bmatrix} \mathbf{v}_2 &= \begin{bmatrix} 1 \\ -1 \end{bmatrix} \end{aligned}$$

which leads to $\mathbf{v}_2 = \begin{bmatrix} 0 \\ 1 \end{bmatrix}$. The second solution is then given by (3.10.11):

$$\begin{aligned} \mathbf{x}_2(t) &= e^{\lambda t}(\mathbf{v}_1 t + \mathbf{v}_2) \\ &= e^{3t}\left(\begin{bmatrix} 1 \\ -1 \end{bmatrix} t + \begin{bmatrix} 0 \\ 1 \end{bmatrix}\right) \end{aligned}$$

Hence, a general solution of the corresponding system is given by (3.10.5):

$$\begin{aligned} \mathbf{x}(t) &= c_1\mathbf{x}_1(t) + c_2\mathbf{x}_2(t) \\ &= c_1 e^{3t}\begin{bmatrix} 1 \\ -1 \end{bmatrix} + c_2 e^{3t}\left(\begin{bmatrix} 1 \\ -1 \end{bmatrix} t + \begin{bmatrix} 0 \\ 1 \end{bmatrix}\right) \end{aligned}$$

THE EIGENVALUE METHOD FOR INHOMOGENEOUS SYSTEMS

The methods of undetermined coefficients (§3.3) and variation of parameters (§3.5) for finding particular solutions of a single inhomogeneous n^{th} order linear differential equation, can both be generalized to inhomogeneous linear systems. Here, we briefly illustrate the extension of the methods of undetermined coefficients and variation of parameters to inhomogeneous systems of first order differential equations with constant coefficients.

The Method of Undetermined Coefficients for Systems

Consider the system

$$\mathbf{x}'(t) = A\mathbf{x}(t) + \mathbf{g}(t) \qquad (3.10.13)$$

where A is the usual $(n \times n)-$ matrix of (constant) coefficients and g is an $(n \times 1)-$ matrix or vector whose entries consist of a linear combination of products of polynomials, sines and cosines and exponential functions. We proceed as in Section 3.3 except that we now use undetermined *vector* coefficients instead of undetermined *scalar* coefficients. To see this, consider the following example.

Example 3.10.5 Solve the system

$$\begin{aligned} x_1' &= 2x_1 + x_2 - 4t \\ x_2' &= x_1 + 2x_2 + 3t + 2 \end{aligned} \qquad (3.10.14)$$

Solution. Since the system is linear, as in the case of the single linear equation (1.4), the general solution splits into two parts(see equation (3.1.1)):

$$\mathbf{x}(t) = \mathbf{x}_c(t) + \mathbf{x}_p(t)$$

where $\mathbf{x}_c$ and $\mathbf{x}_p$ are the complementary and particular solutions, respectively, of the system. In Example 3.10.2, we found that the homogeneous system has general solution

$$\mathbf{x}_c(t) = c_1 \begin{bmatrix} 1 \\ -1 \end{bmatrix} e^t + c_2 \begin{bmatrix} 1 \\ 1 \end{bmatrix} e^{3t}$$

For $\mathbf{x}_p$ try a solution of the form ('generalizing' one of the entries in Table 3.3.1)

$$\mathbf{x}_p(t) = \mathbf{a}t + \mathbf{b} = \begin{bmatrix} a_1 \\ a_2 \end{bmatrix} t + \begin{bmatrix} b_1 \\ b_2 \end{bmatrix} \qquad (3.10.15)$$

Substitute (3.10.15) into the system (3.10.14) to obtain

$$\begin{bmatrix} a_1 \\ a_2 \end{bmatrix} = \begin{bmatrix} 2(a_1t + b_1) + a_2t + b_2 - 4t \\ a_1t + b_1 + 2(a_2t + b_2) + 3t + 2 \end{bmatrix}$$

$$= \begin{bmatrix} (2a_1 + a_2 - 4)\,t + 2b_1 + b_2 \\ (2a_2 + a_1 + 3)\,t + 2 + b_1 + 2b_2 \end{bmatrix}$$

Equating the coefficients of t and the constant terms (in both scalar equations), we obtain

$$\begin{aligned} 2a_1 + a_2 - 4 &= 0 \\ 2b_1 + b_2 &= a_1 \\ (2a_2 + a_1 + 3) &= 0 \\ 2 + b_1 + 2b_2 &= a_2 \end{aligned}$$

Solving, we obtain,

$$b_2 = -\frac{43}{9}, \quad b_1 = \frac{38}{9}, \quad a_2 = -\frac{10}{3}, \quad a_1 = \frac{11}{3}$$

so that

$$\begin{aligned} \mathbf{x}_p\,(t) &= \begin{bmatrix} a_1 \\ a_2 \end{bmatrix} t + \begin{bmatrix} b_1 \\ b_2 \end{bmatrix} \\ &= \begin{bmatrix} \frac{11}{3} \\ -\frac{10}{3} \end{bmatrix} t + \begin{bmatrix} \frac{38}{9} \\ -\frac{43}{9} \end{bmatrix} \end{aligned}$$

and the general solution of the system is

$$\begin{aligned} \mathbf{x}(t) &= \mathbf{x}_c\,(t) + \mathbf{x}_p\,(t) \\ &= c_1 \begin{bmatrix} 1 \\ -1 \end{bmatrix} e^t + c_2 \begin{bmatrix} 1 \\ 1 \end{bmatrix} e^{3t} + \frac{1}{3} \begin{bmatrix} 11 \\ -10 \end{bmatrix} t + \frac{1}{9} \begin{bmatrix} 38 \\ -43 \end{bmatrix} \end{aligned}$$

Variation of Parameters for Systems

The particular solution $\mathbf{x}_p\,(t)$ of Example 3.10.5 could also have been found by variation of parameters as in Section 3.5 by suggesting that, from the expression for $\mathbf{x}_c\,(t)$, the particular solution take the form

$$\mathbf{x}_p\,(t) = c_1\,(t) \begin{bmatrix} 1 \\ -1 \end{bmatrix} e^t + c_2\,(t) \begin{bmatrix} 1 \\ 1 \end{bmatrix} e^{3t} \qquad (3.10.16)$$

where, now, c_1 and c_2 are functions of t to be determined as in Section 3.5, that is, the expression (3.10.16) is substituted into the system (3.10.14) and the functions c_1 and c_2 determined from their derivatives - by integration.

PART 2

5 MIDTERM AND 5 FINAL PRACTICE EXAMINATIONS (WITH DETAILED SOLUTIONS) IN INTRODUCTORY ORDINARY DIFFERENTIAL EQUATIONS

MIDTERM EXAMINATION #1

Time: 60 minutes **Level of Difficulty: 3**

1. (50%) Find the general solution of each of the following differential equations.

   (i) (10%)

$$\frac{dy}{dx} = \frac{xy + x}{y^2 - xy^2}$$

   (ii) (10%)

$$(xy^2)\frac{dy}{dx} = x^3 + y^3$$

   (iii) (10%)

$$x\frac{dy}{dx} = x - y + 2xy$$

   (iv) (10%)

$$\frac{dy}{dt} = \frac{3t^2 + 2y}{y - 2t - 3}$$

   (v) (10%)

$$\frac{dy}{dx} = \cot y$$

2. (20%) Solve the following initial value problem.

$$\frac{dx}{dt} - \frac{1}{t}x = tx^2, \qquad x(1) = 1$$

3. (10%) Find the general solution of

$$D^2(D+2)(D+1)^2(D^2+D+1)^2 y = 0$$

   where $y(x)$ and $D^n \equiv \frac{d^n}{dx^n}$, $n = 1, 2, \ldots$.

4. (20%) One solution of the following differential equation is $y = x$. Use this information, to find a second linearly independent solution and hence the general solution. (You must prove that the solutions obtained are, in fact, linearly independent).

$$2x^2\frac{d^2y}{dx^2} - x\frac{dy}{dx} + y = 0, \qquad x \neq 0$$

MIDTERM EXAMINATION #2

Time: 60 minutes **Level of Difficulty: 3**

1.

(a) (20%) Solve the following initial value problem

$$\frac{d^2y}{dx^2} - \frac{dy}{dx} - 2y = e^{-x} + x, \qquad y(0) = 0, \ y'(0) = 1$$

(b) (10%) If the method of undetermined coefficients is to be used to solve the following differential equation, write down the *form* of the particular solution $y_p(x)$ (**Do not evaluate the coefficients**).

$$\frac{d^3y}{dx^3} + \frac{dy}{dx} = 3 + \frac{3}{2}\cos x + xe^x \sin x$$

2. (15%) Find the orthogonal trajectories associated with the following family of curves with parameter a.

$$yx^2 + 3 = ay$$

3. (10%) Show that the functions

$$x, \quad \cos(\ln x), \quad \sin(\ln x)$$

are linearly independent solutions of the equation

$$x^3\frac{d^3y}{dx^3} + 2x^2\frac{d^2y}{dx^2} + x\frac{dy}{dx} - y = 0, \qquad x > 0$$

Hence or otherwise, write down the general solution.

4. Find the general solution

(i) (10%)

$$\frac{dy}{dx} = \frac{-3x^2e^y}{x^3e^y + \sin y}$$

(ii) (10%)

$$\cos y\frac{dy}{dx} = xe^x$$

(iii) (10%)

$$\frac{dy}{dx} = \frac{y+x}{x}$$

5. (15%) Solve the following initial value problem

$$\frac{dy}{dx} = \frac{x}{x^2\cos y + xy}, \qquad y(0) = 1$$

MIDTERM EXAMINATION #3

Time: 60 minutes **Level of Difficulty: 4**

1. (20%) Consider the pair of differential equations

$$
\begin{aligned}
(x-1)\,y'' - xy' + y &= 0 \qquad \text{(mt3.1)}\\
(x-1)\,y'' - xy' + y &= 1
\end{aligned}
$$

   Given that $y_1(x) = e^x$ solves the homogeneous differential equation (mt3.1), find the general solution of each differential equation.

2. Find the general solution

   (i) (10%)

$$y(x^2+2)dx = (x^3-x)dy$$

   (ii) (10%)

$$\frac{dy}{dx} = \frac{x^4 \cos 3x + 3y}{x}$$

3. (15%) Find the general solution of the equation

$$\frac{d^2y}{dx^2} - 4\frac{dy}{dx} + 4y = 6xe^{2x}$$

4. (15%) Solve the equation

$$\frac{dy}{dx} = \frac{1-xy}{x\,(x-y)}$$

5. (20%) Find the general solution of

$$\frac{d^2y}{dx^2} + 2\frac{dy}{dx} + y = \frac{1}{(e^x-1)^2}$$

6. (10%) The current i at time t in an electrical circuit is modelled by the differential equation

$$\frac{di}{dt} = \frac{V - Ri}{L}$$

   where V is voltage, R is resistance and L is inductance. If V, R and L are positive constants and the current is initially at zero, show that the current (eventually) reaches a steady state of $i = i_c = \dfrac{V}{R}$.

MIDTERM EXAMINATION #4

Time: 60 minutes **Level of Difficulty: 4**

1. (15%) Show that the solution of the initial value problem

$$\left[t\cos^2\left(\frac{y}{t}\right) - y\right]dt = -tdy, \qquad y(1) = \frac{\pi}{4}$$

is given by $\tan\frac{y}{t} = \ln\frac{e}{|t|}$.

2. (20%) Find the orthogonal trajectories corresponding to the family of curves with parameter k described by the equation $y^2 = x^2(1 - kx)$. In particular, find the orthogonal trajectory passing through the point $(1, 1)$.

3. Find the general solution.

(i) (10%)

$$D^2(D^2 - 2D + 2)^2(D^2 - 3D + 2)y = f(x) \qquad \text{(mt4.1)}$$

where $f(x) = 0$, $D^n y \equiv \dfrac{d^n y}{dx^n}$, $n = 1, 2, \ldots$ and y is a function of the independent variable x.

(ii) (10%) $\cos\theta\dfrac{dr}{d\theta} = -\sin\theta\,(1 - 2r)$, where r is a function of θ.

(iii) (15%) $\dfrac{d^2y}{dx^2} + y = \sec x$

4. (10%) Consider the above differential equation (mt4.1). Suppose the right-hand side is replaced by the function $f(x) = 3 + xe^x\sin x + 3e^{4x}$. If a particular solution $y_p(x)$ is sought using the method of undetermined coefficients, write down the *form* of $y_p(x)$ - **do not evaluate the coefficients.**

5. (10%) Solve

$$\frac{dy}{dx} = \frac{-y\cos x - \cos y}{\sin x - x\sin y}$$

given that $y(1) = 0$.

6. (10%) Show that the functions

$$y_1(x) = x, \quad y_2(x) = \frac{1}{x}, \quad y_3(x) = x^2$$

are linearly independent solutions of the equation

$$x^3y''' + x^2y'' - 2xy' + 2y = 0, \quad x \neq 0$$

Hence or otherwise, find the general solution.

MIDTERM EXAMINATION #5

Time: 90 minutes **Level of Difficulty: 5**

1. Find the general solution.

   (i) (5%) $\dfrac{1}{e^y}\dfrac{dy}{dt} = 1 + e^t - e^{-y} - e^{t-y}$

   (ii) (15%) $\dfrac{dy}{dx} = \dfrac{6y^2}{x\left(2x^3 + y\right)}$

2. (10%) Find all possible functions $g(x)$ such that the differential equation
$$\frac{dy}{dx} = -\frac{y \sin x}{g(x)}$$
is exact. Solve the differential equation for these functions $g(x)$.

3. (10%) Solve the initial value problem
$$(t-y)dt + (3t+y)dy = 0, \quad y\left(1\right) = 0$$

4. (15%) It is known that the differential equation
$$x^3 y'' + xy' - y = 0$$
has a solution of the form $y = x^n$, for some positive integer n. Find such a solution and hence find the general solution of the inhomogeneous equation
$$x^3 y'' + xy' - y = 1$$

5. **(a)** (5%) Is every separable equation exact ? Justify your answer with either a proof or a counter-example.

   (b) (5%) If the Wronskian of two functions is identically zero, is it the case that the two functions must be linearly dependent ? Justify your answer with either a proof or a counter-example.

6. (10%) Solve the boundary value problem
$$\frac{d^2x}{dt^2} - 2\frac{dx}{dt} + 2x = e^t \cos 2t, \quad x'\left(0\right) = 0, \quad x\left(0\right) = 1$$

7. (10%) Investigate the solution of the boundary value problem
$$\frac{d^2y}{dx^2} + y = x^3, \quad y\left(0\right) = 0, \quad y\left(\pi\right) = 0$$

8. (15%) Use a suitable substitution to solve
$$\left[\left(\frac{x}{y}\right)^2 \left(1 + e^{yx^{-1}}\right)^{-1} + \frac{y}{x}\right] dx - dy = 0, \quad y\left(1\right) = 0$$

FINAL EXAMINATION #1

Time: 2 Hours **Level of Difficulty: 3**

1. (20%) Solve the following differential equation

$$x^2\frac{d^2y}{dx^2}+x\frac{dy}{dx}-y=x^2\ln x, \quad x>0$$

2. (10%) Use the method of *reduction of order* to find the general solution of the differential equation

$$\frac{d^2y}{dx^2}-4\frac{dy}{dx}+4y=\frac{e^{2x}}{x^2}, \quad x>0$$

3. (10%) Determine the appropriate form for a particular solution of the differential equation

$$(D-1)^4\left(D^2+16\right)^2y=xe^x+x\cos 4x$$

when using the method of undetermined coefficients. **Do not evaluate the coefficients.** Note that, here, $D^n\equiv\dfrac{d^n}{dx^n}$.

4. Find the Laplace transform of each of the following functions.

(i) (10%)

$$g(t)=\begin{cases}\cos t, & 0<t<\pi,\\ 0, & \pi<t<2\pi,\end{cases}$$

$g(t)=g(t+2\pi),\ t\geq 0.$

(ii) (5%)

$$h(t)=e^t\cos 2t+t\sin t$$

5. (15%) Solve the following initial value problem using the method of Laplace transforms.

$$\begin{aligned}\frac{d^2x}{dt^2}+x &= \begin{cases}2, & 0\leq t<\frac{\pi}{2},\\ 0, & t\geq\frac{\pi}{2},\end{cases}\\ x(0) &= 0,\quad \frac{dx}{dt}(0)=2\end{aligned}$$

6. (10%) Use the convolution theorem and the method of Laplace transforms to solve
$$\frac{d^2y}{dx^2} + 6\frac{dy}{dx} + 9y = H(x), \quad y(0) = 0, \quad y'(0) = 0$$
where $H(x)$ is a known function of x.

7. (20%) Find the general solution in a series about $x = 0$. Give a region of validity for your solution (you must justify your conclusion).
$$\left(4 - x^2\right) y'' - 2xy' + 2y = 0$$

FINAL EXAMINATION #2

Time: 2 Hours **Level of Difficulty: 3**

1. (10%) Solve the following nonlinear ordinary differential equation.

$$y\frac{d^2y}{dx^2} - \left(\frac{dy}{dx}\right)^2 = 0, \quad y > 0$$

2. (15%) For the following homogeneous differential equation find a solution of the form x^n, where n is a positive integer.

$$(x^2 - 1)\frac{d^2y}{dx^2} - 2x\frac{dy}{dx} + 2y = 0$$

Use this solution to find the general solution of the following corresponding inhomogeneous equation

$$(x^2 - 1)\frac{d^2y}{dx^2} - 2x\frac{dy}{dx} + 2y = \left(x^2 - 1\right)^2$$

3.

(i) (10%) Find

$$L^{-1}\left[\frac{5s + 3}{s^2 + 4s + 5}\right]$$

(ii) (10%) Determine

$$L^{-1}\left[\frac{s}{(s+3)^5\,(s^2 + 16)}\right]$$

in the form of an integral (Do not evaluate the integral).

4.

(a) (10%) Find the general solution of the equation

$$y'' - 6y' + 13y = 15\cos 2x \tag{F2.1}$$

(b) (5%) If the left-hand side of (F2.1) is replaced by $y'' + 4y$, what would be the corresponding form of particular solution y_p if the method of undetermined coefficients is used to find the general solution?

5. (15%) Solve the following initial value problem using the method of Laplace transforms.

$$\frac{d^2y}{dx^2} + 4\frac{dy}{dx} + 4y = e^{-2x}, \quad y(0) = 0, \quad y'(0) = 1$$

6. (25%) Find the general solution of the following differential equation in terms of series centered at $x = 0$.

$$2x^2(1-x)\frac{d^2y}{dx^2} - x(1+x)\frac{dy}{dx} + (1+x)y = 0, \qquad x > 0$$

FINAL EXAMINATION #3

Time: 2 Hours **Level of Difficulty: 4**

1. (15%) Solve the boundary value problem

$$\frac{d^2y}{dt^2}\cos\left(\frac{dy}{dt}\right) = \cos t, \qquad -\frac{\pi}{2} \le t \le \frac{\pi}{2}, \qquad y(1) = 1, \ \frac{dy}{dt}(0) = 0$$

Comment on the case $t \notin \left[-\frac{\pi}{2}, \frac{\pi}{2}\right]$.

2. (10%) Classify all singular points (excluding the point at infinity) of the following ordinary differential equation

$$\left(1 - x^2\right)^2 x\frac{d^2y}{dx^2} - 2x(1+x)\frac{dy}{dx} + \beta(\beta - 1)y = 0$$

where β is a real number.

(i) (10%) Given that $y = x$ is one solution of the homogeneous equation

$$\left(x^2 + 1\right)y'' - 2xy' + 2y = 0$$

find its general solution.

(ii) (15%) Find the general solution of the inhomogeneous equation

$$\frac{d^2y}{dx^2} - 7\frac{dy}{dx} + 12y = \sin\left(e^{-3x}\right)$$

3. (10%) Use the method of Laplace transforms to solve the following initial value problem.

$$\begin{aligned} \frac{dy}{dt} + x(t) &= A(t) \\ \frac{dx}{dt} - y(t) &= B(t) \\ x(0) &= 0, \ y(0) = 1 \end{aligned}$$

where A and B are given functions of t. (You may leave your answer in terms of integrals).

(i) (20%) Find the general solution in series about $x = 0$.

$$x\frac{d^2y}{dx^2} + 2\frac{dy}{dx} + 11xy = 0, \qquad x > 0$$

(ii) (5%) Write the general solution from 5(i) above in terms of known functions.

4. (15%) Find the general solution of the following system of equations.

$$\begin{aligned}
\frac{dx}{dt} &= 3x + 2y + 2z \\
\frac{dy}{dt} &= -5x - 4y - 2z \\
\frac{dz}{dt} &= 5x + 5y + 3z
\end{aligned}$$

FINAL EXAMINATION #4

Time: 2 Hours **Level of Difficulty: 4**

1. (10%) Solve the initial value problem

$$\frac{dy}{dx} = -\frac{(x+1)\tan y}{x\sec^2 y}, \qquad y(1) = \frac{\pi}{4}$$

2. (15%) Use the method of Laplace transforms to solve the following initial value problem.

$$\frac{d^2y}{dt^2} + y(t) = H(t), \qquad y(0) = 1, \ \frac{dy}{dt}(0) = 1$$

where $H(t)$ is given by

$$H(t) = \begin{cases} t, & 0 \le t < 1, \\ 0, & t \ge 1 \end{cases}$$

3. (10%) Find the general solution of the equation

$$(e^x + 1)\frac{d^2y}{dx^2} - (e^x + 1)\frac{dy}{dx} = 1$$

4. **(a)** (10%)Classify all singular points *and* the point at infinity of the differential equation

$$\left(1 - t^2\right)\frac{d^2y}{dt^2} - \frac{2t}{(1+t)^2}\frac{dy}{dt} + n(n+1)y = 0$$

where n is a positive integer.

(b) (20%) Find the general solution of the following differential equation in terms of power series about $x = 1$. State the region of validity of your solution (you must justify your conclusion).

$$y'' - (x-1)y' + 5y = 0$$

Here, y is a function of x.

5. Find the following.

   (i) (5%) $L\left[t^2 \cos t\right]$

   (ii) (7%) $L[F(t)]$ where

$$F(t) = \begin{cases} 0, & 0 \le t < 2, \\ \cos \pi t, & 2 \le t < 3, \\ 0, & t \ge 3 \end{cases}$$

6. (8%) Use the derivative of the function

$$f(s) = \ln\left(2 + \frac{3}{s}\right), \quad s > 0$$

   to find its inverse Laplace transform.

7. (15%) Find a series solution of the following differential equation near $x = 0$. What form should the other linearly independent solution take (you must justify your conclusion) ?

$$x\frac{d^2y}{dx^2} + (1 - x)\frac{dy}{dx} + \alpha y = 0, \quad x > 0, \quad \alpha \text{ is a positive integer}$$

FINAL EXAMINATION #5

Time: 2 Hours **Level of Difficulty: 5**

1. (25%) Find the general solution, valid near $x = 0$, of Bessel's equation of order 1

$$x^2 \frac{d^2y}{dx^2} + x\frac{dy}{dx} + \left(x^2 - 1\right) y = 0, \qquad x > 0$$

2. (15%) Use the Laplace transform method to solve the following initial value problem.

$$\begin{aligned} \frac{d^2x}{dt^2} + 3\frac{dx}{dt} &= H(t) \\ x(0) &= \frac{dx}{dt}(0) = 0 \end{aligned}$$

where

$$H(t) = \begin{cases} \cos 3t, & 0 \le t < 3\pi, \\ 1 + \cos 3t, & t \ge 3\pi \end{cases}$$

3. (15%) Use the *method of undetermined coefficients* to find the general solution of

$$\frac{d^2y}{dx^2} + 4y = 2\sin^2 x$$

Which other method is suitable in this case ? Justify your conclusion.

4. (10%) Find $L[|\sin kt|]$ where k is a non-zero integer.

5. (10%) Consider Bessel's equation of order zero

$$x\frac{d^2y}{dx^2} + \frac{dy}{dx} + xy = 0$$

Using the method of series, it can be shown that one (continuously differentiable) solution of this equation is the *Bessel function of order zero*, denoted by $J_0(x)$. Find a second linearly independent solution (you may leave your solution in terms of an integral and the function $J_0(x)$).

6. (15%) Find the general solution of the system

$$\begin{aligned} \frac{dx}{dt} &= 3x + 2y + z \\ \frac{dy}{dt} &= -x - z \\ \frac{dz}{dt} &= x + y + 2z \end{aligned}$$

7. (10%) Solve the following initial value problem.

$$
\begin{aligned}
e^{2x}\frac{d^2x}{dt^2} &= -1, \qquad t > 0,\ \frac{dx}{dt} > 0 \\
x(0) &= 0,\ \frac{dx}{dt}(0) = 1
\end{aligned}
$$

MIDTERM EXAMINATION #1
SOLUTIONS

In what follows, c_i, $i = 1, 2, ...$ will denote arbitrary constants. Rules of logarithms and exponentials, a summary of the main techniques of integration as well as a *table of integrals* can be found in the Appendix.

1. (50%) Find the general solution of each of the following differential equations.

(i) (10%)

$$\frac{dy}{dx} = \frac{xy + x}{y^2 - xy^2}$$

(ii) (10%)

$$(xy^2)\frac{dy}{dx} = x^3 + y^3$$

(iii) (10%)

$$x\frac{dy}{dx} = x - y + 2xy$$

(iv) (10%)

$$\frac{dy}{dt} = \frac{3t^2 + 2y}{y - 2t - 3}$$

(v) (10%)

$$\frac{dy}{dx} = \cot y$$

Solution. The key to solving first order ordinary differential equations is in *identification* and *classification*: try to identify the given equation as belonging to one of the classes discussed in Chapter 2. This is done by a systematic process of elimination: we examine the differential equation for the characteristics of a particular 'class' discussed in Chapter 2 - if these characteristics are absent, we move to the next class and continue in this fashion until the appropriate class has been identified. Next, we apply the corresponding solution technique to reduce the differential equation problem to a calculus problem. The general solution is then obtained using one or more of the various different techniques of integration (see the Appendix) and algebra.

Question 1.(i) Simplify the right-hand side in order to identify/classify the differential equation.

$$\begin{aligned} \frac{dy}{dx} &= \frac{xy+x}{y^2-xy^2} = \frac{x(y+1)}{y^2(1-x)} \\ &= \left[\frac{x}{(1-x)}\right]\left[\frac{(y+1)}{y^2}\right] \end{aligned}$$

The equation is *separable* (see Section 2.1). Following the solution procedure of Section 2.1, we have

$$\int \frac{y^2}{1+y}dy = \int \frac{x}{1-x}dx$$

These integrals are found by dividing the integrands and obtaining proper rational forms.

$$\begin{aligned} \int \left[y-1+\frac{1}{1+y}\right] dy &= \int \left[-1+\frac{1}{1-x}\right] dx \\ \frac{y^2}{2}-y+\ln|1+y| &= -x-\ln|1-x|+c_1 \\ \frac{y^2}{2}-y+\ln|1+y| &= -(x+\ln A|1-x|), \qquad c_1=-\ln A, \quad A>0 \\ \frac{y^2}{2}-y+x &= -\ln A|1-x|\,|1+y| \end{aligned}$$

The general solution is obtained in 'implicit' form.

Question 1.(ii) Try to classify the differential equation. Again, we examine the right-hand side of the differential equation.

$$\begin{aligned} (xy^2)\frac{dy}{dx} &= x^3+y^3 \\ \frac{dy}{dx} &= \frac{x^3+y^3}{xy^2} = f(x,y), \text{ say} \end{aligned}$$

The equation is not separable but may be *homogeneous* (see Section 2.2). In fact, following the procedure in Section 2.2,

$$\begin{aligned} f(\lambda x,\lambda y) &= \frac{(\lambda x)^3+(\lambda y)^3}{(\lambda x)(\lambda y)^2} \\ &= \frac{x^3+y^3}{xy^2} \\ &= f(x,y) \end{aligned}$$

This confirms that the differential equation is indeed homogeneous. Apply the appropriate solution procedure for homogeneous equations (Section 2.2): let $y=vx$ so that

$$\frac{dy}{dx} = v + x\frac{dv}{dx}$$

The differential equation then becomes

$$\begin{aligned}
v + x\frac{dv}{dx} &= \frac{x^3 + v^3x^3}{xv^2x^2} = \frac{1+v^3}{v^2} \\
x\frac{dv}{dx} &= \frac{1+v^3-v^3}{v^2} = \frac{1}{v^2} \\
\int v^2 dv &= \int \frac{dx}{x} \\
\frac{v^3}{3} &= \ln|x| + c_1 = \ln A\,|x|\,, \qquad c_1 = \ln A, \quad A > 0
\end{aligned}$$

Let $v = \dfrac{y}{x}$ to obtain the general solution.

$$\begin{aligned}
\frac{y^3}{3x^3} &= \ln A\,|x| \\
y^3 &= 3x^3 \ln A\,|x|
\end{aligned}$$

Question 1.(iii) Try to classify the equation. Once the 'type' is identified, apply the corresponding technique. The equation is neither separable nor homogeneous. However, we note that y appears only to the power one. This suggests that the equation may be linear in y. In fact,

$$\begin{aligned}
x\frac{dy}{dx} &= x - y + 2xy \\
x\frac{dy}{dx} + (1-2x)\,y &= x \\
\frac{dy}{dx} + \left(\frac{1-2x}{x}\right) y &= 1
\end{aligned}$$

The equation is *linear in* y (see Section 2.5). Following the solution procedure of Section 2.5, we find an integrating factor.

$$\begin{aligned}
\mu(x) &= \exp\left(\int \left(\frac{1-2x}{x}\right) dx\right) \\
&= \exp\left(\int \left(\frac{1}{x} - 2\right) dx\right) \\
&= \exp\left(\ln|x| - 2x\right) \\
&= |x|\, e^{-2x}
\end{aligned}$$

Choose

$$\mu(x) = xe^{-2x}$$

Multiplying both sides of the differential equation by $\mu(x)$, we obtain

$$xe^{-2x}\left(\frac{dy}{dx}+\left(\frac{1-2x}{x}\right)y\right)=xe^{-2x}\cdot 1$$

As in (2.5.3), this can be written in the form

$$\frac{d}{dx}\left(yxe^{-2x}\right)=xe^{-2x}$$

Integrating both sides with respect to x, we obtain

$$\begin{aligned} yxe^{-2x} &= \underbrace{\int xe^{-2x}dx}_{\text{Integration by Parts - Appendix}} \\ &= -\frac{xe^{-2x}}{2}+\frac{1}{2}\int e^{-2x}dx \\ &= -\frac{xe^{-2x}}{2}-\frac{1}{4}e^{-2x}+c_1 \end{aligned}$$

Hence, the general solution is given by

$$y=-\frac{1}{2}-\frac{1}{4x}+\frac{c_1e^{2x}}{x}$$

Question 1.(iv) The equation is neither separable, nor homogeneous, nor linear in y (or t) but may be exact. To apply the test for exactness (see (2.3.5) in Section 2.3) we first write the differential equation in 'standard form' (2.3.1).

$$\begin{aligned} \frac{dy}{dt} &= \frac{3t^2+2y}{y-2t-3} \\ \underbrace{(3t^2+2y)}_{M(t,y)}\,dt+\underbrace{(3-y+2t)}_{N(t,y)}\,dy &= 0 \end{aligned}$$

Now,

$$\frac{\partial M}{\partial y}=\frac{\partial}{\partial y}\left(3t^2+2y\right)=2=\frac{\partial N}{\partial t}$$

Hence, the equation is *exact* (see Section 2.3) and from the solution procedure in Section 2.3, there must exist a function $F(t,x)$ satisfying (see (2.3.2))

$$\frac{\partial F}{\partial t} = M(t,y)=3t^2+2y \qquad \text{(mt1.1)}$$

$$\frac{\partial F}{\partial y} = N(t,y)=(3-y+2t) \qquad \text{(mt1.2)}$$

such that the general solution of the differential equation is given by

$$F(t,y)=c_1$$

To find the function F, we integrate either of the equations (mt1.1) or (mt1.2) and use the remaining equation to evaluate the arbitrary function. From (mt1.2) (the 'easier' of the two to integrate)

$$F(t, y) = 3y - \frac{y^2}{2} + 2ty + f(t) \tag{mt1.3}$$

where f is an arbitrary function of t. The F given by (mt1.3) must also satisfy (mt1.1) - this will determine $f(t)$. From (mt1.3)

$$\frac{\partial F}{\partial t} = 2y + f'(t)$$

Comparing this with (mt1.1) we see that $f'(t) = 3t^2$ so that $f(t) = t^3 + c_2$. From (mt1.3), we have that the function F is given by

$$F(t, y) = 3y - \frac{y^2}{2} + 2ty + t^3 + c_2$$

Finally, the general solution of the differential equation is given by

$$\begin{aligned} F(t, y) &= c_1 \\ 3y - \frac{y^2}{2} + 2ty + t^3 + c_2 &= c_1 \\ 3y - \frac{y^2}{2} + 2ty + t^3 &= c_3, \qquad c_3 = c_1 - c_2 \\ y\left(3 - \frac{y}{2} + 2t\right) + t^3 &= c_3 \end{aligned}$$

Question 1.(v) The differential equation

$$\frac{dy}{dx} = \cot y$$

is separable (see Section 2.1). Hence,

$$\begin{aligned} \int \frac{dy}{\cot y} &= \int dx \\ \int \tan y dy &= x + c_1 \\ \underbrace{\int \frac{\sin y}{\cos y} dy}_{\text{Let } u = \cos y} &= x + c_1 \\ \ln|\cos y| &= x + c_1 \\ |\cos y| &= \exp(x + c_1) = c_2 e^x, \qquad c_2 = e^{c_1} \end{aligned}$$

The general solution is therefore

$$\cos y = c_3 e^x, \qquad c_3 = \pm c_2$$

Question 2. (20%) Solve the following initial value problem.

$$\frac{dx}{dt} - \frac{1}{t}x = tx^2, \qquad x(1) = 1$$

Solution. The differential equation

$$\frac{dx}{dt} - \frac{1}{t}x = tx^2$$

would be linear in $x(t)$ if not for the term in x^2 on the right-hand side. In fact, the equation is of the *Bernoulli* type (see Section 2.6) with $n = 2$. Following the appropriate procedure, let $v = x^{1-2} = x^{-1}$ or $x = v^{-1}$. Then,

$$\frac{dx}{dt} = \frac{dx}{dv}\frac{dv}{dt} = -\frac{1}{v^2}\frac{dv}{dt}$$

and the differential equation becomes

$$-\frac{1}{v^2}\frac{dv}{dt} - \frac{1}{t}\frac{1}{v} = \frac{t}{v^2}$$

or

$$\frac{dv}{dt} + \frac{v}{t} = -t$$

which is *linear in* v (see Section 2.5). An integrating factor for this equation is given by

$$\begin{aligned} \mu(t) &= \exp\left(\int \frac{dt}{t}\right) \\ &= \exp(\ln|t|) \end{aligned}$$

Choose $\mu(t) = t$. Hence, we have

$$\begin{aligned} \frac{d}{dt}(tv) &= -t^2 \\ tv &= -\frac{t^3}{3} + c_1 \\ v &= -\frac{t^2}{3} + c_1 t^{-1} \end{aligned}$$

Next, $x = v^{-1}$ so that the general solution of the differential equation is given by

$$\begin{aligned} \frac{1}{x(t)} &= -\frac{t^2}{3} + \frac{c_1}{t} = \frac{-t^3 + 3c_1}{3t} \\ x(t) &= \frac{3t}{c_2 - t^3}, \qquad c_2 = 3c_1 \end{aligned}$$

Applying the initial condition $x(1) = 1$, we obtain

$$\begin{aligned} 1 &= \frac{3}{c_2 - 1} \\ c_2 &= 4 \end{aligned}$$

so that the solution of the initial value problem is given by

$$x(t) = \frac{3t}{4 - t^3}$$

Question 3.(10%) Find the general solution of

$$D^2(D+2)(D+1)^2(D^2+D+1)^2 y = 0$$

where $y(x)$ and $D^n \equiv \dfrac{d^n}{dx^n}$, $n = 1, 2, \ldots$.

Solution. The differential equation is a $9th$-order linear homogeneous ordinary differential equation with constant coefficients (see Section 3.2). We use Table 3.2.1 to 'piece together' the general solution.

$$\begin{aligned} y(x) &= \underbrace{c_1 + c_2 x}_{\text{For } D^2} + \underbrace{c_3 e^{-2x}}_{\text{For } D+2} + \underbrace{(c_4 + c_5 x)\, e^{-x}}_{\text{For } (D+1)^2} + \\ &\quad \underbrace{+\, e^{-\frac{1}{2}x}\left[(c_6 + c_7 x)\cos\frac{\sqrt{3}}{2}x + (c_8 + c_9 x)\sin\frac{\sqrt{3}}{2}x\right]}_{\text{For } (D^2+D+1)^2 = \left([D-(-\frac{1}{2}+i\frac{\sqrt{3}}{2})][D-(-\frac{1}{2}-i\frac{\sqrt{3}}{2})]\right)^2} \end{aligned}$$

Note that we have 9 arbitrary constants - as expected from a 9th order equation.

Question 4.(20%)One solution of the following differential equation is $y = x$. Use this information, to find a second linearly independent solution and hence the general solution. (You must prove that the solutions obtained are, in fact, linearly independent).

$$2x^2\frac{d^2y}{dx^2} - x\frac{dy}{dx} + y = 0, \qquad x \neq 0$$

Solution. We are given one solution and we are required to find a second linearly independent solution - this information suggests that we use *reduction of order* (see Section 3.4). Let $y = vx$. Then

$$\begin{aligned} \frac{dy}{dx} &= v + x\frac{dv}{dx} \\ \frac{d^2y}{dx^2} &= \frac{dv}{dx} + \frac{dv}{dx} + x\frac{d^2v}{dx^2} \\ &= 2\frac{dv}{dx} + x\frac{d^2v}{dx^2} \end{aligned}$$

Substituting these relations into the differential equation we obtain

$$2x^2\left(2\frac{dv}{dx} + x\frac{d^2v}{dx^2}\right) - x\left(v + x\frac{dv}{dx}\right) + vx = 0$$

At this stage, we expect all the v's to disappear. In fact, simplifying the equation, we obtain

$$2x^3\frac{d^2v}{dx^2} + 3x^2\frac{dv}{dx} = 0$$

Let $w = \dfrac{dv}{dx}$.

$$\frac{dw}{dx} + \frac{3}{2x}w = 0$$

This equation is *linear in* w (see Section 2.5) and has integrating factor $\mu(x) = x^{\frac{3}{2}}$. Hence,

$$\begin{aligned} \frac{d}{dx}\left(x^{\frac{3}{2}}w\right) &= 0 \\ x^{\frac{3}{2}}w &= c_1 \\ w &= c_1x^{-\frac{3}{2}} \end{aligned}$$

Thus,

$$v(x) = \int w(x)dx = c_1 2x^{-\frac{1}{2}} + c_2$$

and the general solution of the differential equation is given by

$$\begin{aligned} y &= vx = c_1 2x^{\frac{1}{2}} + c_2x \\ &= c_3x^{\frac{1}{2}} + c_2x, \qquad c_3 = 2c_1 \end{aligned}$$

The "given" solution $y_1 = x$ is 'attached' to the constant c_2. The second linearly independent solution is attached to the constant c_3 and is therefore $y_2 = x^{\frac{1}{2}}$. To check that these solutions are indeed linearly independent, we use the Wronskian (see Section 3.1).

$$\begin{aligned} W &= \begin{vmatrix} x^{\frac{1}{2}} & x \\ \frac{1}{2}x^{-\frac{1}{2}} & 1 \end{vmatrix} \\ &= x^{\frac{1}{2}} - \frac{1}{2}x^{\frac{1}{2}} \\ &= \frac{1}{2}x^{\frac{1}{2}} \\ &\neq 0, \quad x \neq 0 \end{aligned}$$

Hence, by Note 3.1.1(b), the solutions $\left\{x^{\frac{1}{2}},\ x\right\}$ are indeed linearly independent.

MIDTERM EXAMINATION #2
SOLUTIONS

In what follows, c_i, $i = 1, 2, ...$ will denote arbitrary constants. Rules of logarithms and exponentials, a summary of the main techniques of integration as well as a *table of integrals* can be found in the Appendix.

1.

(a) (20%) Solve the following initial value problem

$$\frac{d^2y}{dx^2} - \frac{dy}{dx} - 2y = e^{-x} + x, \qquad y(0) = 0, \;\; y'(0) = 1$$

(b) (10%) If the method of undetermined coefficients is to be used to solve the following differential equation, write down the *form* of the particular solution $y_p(x)$ (**Do not evaluate the coefficients**).

$$\frac{d^3y}{dx^3} + \frac{dy}{dx} = 3 + \frac{3}{2}\cos x + xe^x \sin x$$

Solution

Question 1(a) The differential equation

$$\frac{d^2y}{dx^2} - \frac{dy}{dx} - 2y = e^{-x} + x$$

is linear, inhomogeneous with constant coefficients. We use the method of undetermined coefficients (see Sections 3.2 and 3.3). The general solution decomposes into two parts:

$$y(x) = y_c(x) + y_p(x)$$

For the complementary solution $y_c(x)$, the characteristic equation is given by (3.2.2):

$$\begin{aligned} m^2 - m - 2 &= 0 \\ (m-2)(m+1) &= 0 \\ m &= -1, 2 \end{aligned}$$

Hence, from Table 3.2.1, the complementary solution is given by

$$y_c(x) = c_1 e^{-x} + c_2 e^{2x}$$

For the particular solution $y_p(x)$, from Table 3.3.1, try

$$y_p(x) = A \underset{\substack{\uparrow \\ (*)}}{x} e^{-x} + Cx + E$$

Here, A, C and E are definite constants to be determined. (*Note that it is necessary to multiply the 'e^{-x}'- term in $y_p(x)$ by 'x' to avoid terms in $y_p(x)$ common to both y_c and y_p - see Section 3.3). To find A, C and E, we substitute this y_p into the differential equation

$$\begin{aligned} y_p(x) &= Axe^{-x} + Cx + E \\ y_p'(x) &= Ae^{-x} - Axe^{-x} + C \\ y_p''(x) &= -2Ae^{-x} + Axe^{-x} \end{aligned}$$

The differential equation becomes

$$\begin{aligned} \frac{d^2y_p}{dx^2} - \frac{dy_p}{dx} - 2y_p &= e^{-x} + x \\ \left(-2Ae^{-x} + Axe^{-x}\right) - \left(Ae^{-x} - Axe^{-x} + C\right) - 2\left(Axe^{-x} + Cx + E\right) &= e^{-x} + x \\ -3Ae^{-x} - 2Cx - 2E - C &= e^{-x} + x \end{aligned}$$

Comparing coefficents of e^{-x} and powers of x, we obtain

$$\begin{aligned} -3A &= 1 \Rightarrow A = -\frac{1}{3} \\ -2C &= 1 \Rightarrow C = -\frac{1}{2} \\ -2E - C &= 0 \Rightarrow E = \frac{1}{4} \end{aligned}$$

Hence,

$$y_p(x) = -\frac{1}{3}xe^{-x} - \frac{1}{2}x + \frac{1}{4}$$

and the general solution of the differential equation is given by

$$\begin{aligned} y(x) &= y_c(x) + y_p(x) \\ &= c_1e^{-x} + c_2e^{2x} - \frac{1}{3}xe^{-x} - \frac{1}{2}x + \frac{1}{4} \end{aligned}$$

Finally,

$$\begin{aligned} y(0) &= 0 \Rightarrow c_1 + c_2 + \frac{1}{4} = 0 \\ y'(0) &= 1 \Rightarrow -c_1 + 2c_2 - \frac{1}{3} - \frac{1}{2} = 1 \end{aligned}$$

Simplifying,

$$\begin{aligned} c_1 + c_2 &= -\frac{1}{4} \\ -c_1 + 2c_2 &= \frac{11}{6} \end{aligned}$$

Using elimination or Cramer's rule to solve this system, we obtain

$$c_2 = \frac{19}{36}, \qquad c_1 = -\frac{7}{9}$$

Hence, the solution of the initial value problem is given by

$$y(x) = -\frac{7}{9}e^{-x} + \frac{19}{36}e^{2x} - \frac{1}{3}xe^{-x} - \frac{1}{2}x + \frac{1}{4}$$

Question 1(b) The differential equation

$$\frac{d^3y}{dx^3} + \frac{dy}{dx} = 3 + \frac{3}{2}\cos x + xe^x \sin x$$

is linear. Hence, its general solution is given by

$$y(x) = y_c(x) + y_p(x)$$

From Table 3.3.1, the specific form of $y_p(x)$ will depend on the complementary solution $y_c(x)$. For $y_c(x)$, the characteristic equation is given by

$$\begin{aligned} m^3 + m &= 0 \\ m\left(m^2 + 1\right) &= 0 \\ m &= 0, \qquad m = \pm i \end{aligned}$$

Hence, from Table 3.2.1,

$$y_c(x) = c_1 + c_2 \cos x + c_3 \sin x$$

For $y_p(x)$, from Table 3.3.1, try

$$y_p(x) = A \underset{\underset{(*)}{\uparrow}}{x} + \underset{\underset{(*)}{\uparrow}}{x} \left(B\cos x + C\sin x\right) + e^x\left[(Ex + F)\sin x + (Gx + H)\cos x\right]$$

where, at $(*)$, it is necessary to multiply by 'x' to avoid terms in $y_p(x)$ common to both y_c and y_p (see Section 3.3). Again, A, B, C, E, F, G and H are definite constants to be evaluated.

Question 2.(15%) Find the orthogonal trajectories associated with the following family of curves with parameter a.

$$yx^2 + 3 = ay \tag{mt2.1}$$

Solution. First find the differential equation associated with the given family of curves. To do this, differentiate (mt2.1) implicitly with respect to x.

$$\begin{aligned} yx^2 + 3 &= ay \\ 2xy + y'x^2 + 0 &= ay' \\ y'\left(x^2 - a\right) + 2xy &= 0 \end{aligned}$$

However, from (mt2.1),

$$a = \frac{yx^2 + 3}{y}$$

Hence, the differential equation for the family of curves is given by

$$\begin{aligned} y'\left(x^2 - \left[\frac{yx^2 + 3}{y}\right]\right) + 2xy &= 0 \\ y'\left(x^2 - x^2 - \frac{3}{y}\right) + 2xy &= 0 \\ y' &= \frac{dy}{dx} = \frac{2xy^2}{3} \end{aligned}$$

The differential equation for the orthogonal trajectories is given by (see Section 2.8)

$$\frac{dy}{dx} = -\frac{3}{2xy^2}$$

which is *separable* (Section 2.1). Hence,

$$\begin{aligned} \int y^2 dy &= \int -\frac{3}{2x} dx \\ \frac{y^3}{3} &= -\frac{3}{2}\ln|x| + c_1 \end{aligned}$$

Now let $c_1 = -\frac{3}{2}\ln A$, $A > 0$ to obtain

$$y^3 = -\frac{9}{2}\ln A|x|, \qquad A > 0$$

which are, therefore, the required orthogonal trajectories.

Question 3.(10%) Show that the functions

$$x, \quad \cos(\ln x), \quad \sin(\ln x)$$

are linearly independent solutions of the equation

$$x^3\frac{d^3y}{dx^3} + 2x^2\frac{d^2y}{dx^2} + x\frac{dy}{dx} - y = 0, \qquad x > 0$$

Hence or otherwise, write down the general solution.

Solution. Substituting, in turn, each of x, $\cos(\ln x)$ and $\sin(\ln x)$ into the differential equation, we find that each satisfies the differential equation identically. Hence, they are indeed solutions of the differential equation. The fact that they are linearly independent can be shown by calculating the Wronskian (Section 3.1) and showing that it is nonzero i.e.

$$\begin{vmatrix} x & \cos(\ln x) & \sin(\ln x) \\ 1 & -\frac{\sin(\ln x)}{x} & \frac{\cos(\ln x)}{x} \\ 0 & -\frac{[\cos(\ln x)-\sin(\ln x)]}{x^2} & -\frac{[\sin(\ln x)+\cos(\ln x)]}{x^2} \end{vmatrix} = \frac{2}{x^2} \neq 0$$

The differential equation is of third order. Hence, its general solution is given by a linear combination of three linearly independent solutions (see Sections 3.1 and 3.2):

$$y(x) = c_1x + c_2\cos(\ln x) + c_3\sin(\ln x)$$

4. Find the general solution

(i) (10%)

$$\frac{dy}{dx} = \frac{-3x^2e^y}{x^3e^y + \sin y}$$

(ii) (10%)

$$\cos y\frac{dy}{dx} = xe^x$$

(iii) (10%)

$$\frac{dy}{dx} = \frac{y + x}{x}$$

Solution

Question 4.(i) The equation

$$\frac{dy}{dx} = \frac{-3x^2e^y}{x^3e^y + \sin y}$$

is neither separable, nor homogeneous, nor linear in y (or x) but may be exact. To apply the test for exactness (see (2.3.5) in Section 2.3) we first write the differential equation in 'standard form' (2.3.1).

$$\begin{aligned} \frac{dy}{dx} &= \frac{-3x^2e^y}{x^3e^y + \sin y} \\ \underbrace{\left(3x^2e^y\right)}_{M(x,y)} dx + \underbrace{\left(x^3e^y + \sin y\right)}_{N(x,y)} dy &= 0 \end{aligned}$$

Now,

$$\frac{\partial M}{\partial y} = \frac{\partial}{\partial y}\left(3x^2e^y\right) = 3x^2e^y = \frac{\partial N}{\partial x}$$

Hence, the equation is *exact* (see Section 2.3) and from the solution procedure in Section 2.3, there must exist a function $F(x, y)$ satisfying (see (2.3.2))

$$\frac{\partial F}{\partial x} = M(x,y) = 3x^2e^y \qquad \text{(mt2.2)}$$

$$\frac{\partial F}{\partial y} = N(x,y) = x^3e^y + \sin y \qquad \text{(mt2.3)}$$

such that the general solution of the differential equation is given by

$$F(x,y) = c_1$$

To find the function F, we integrate either of the equations (mt2.2) or (mt2.3) and use the remaining equation to evaluate the arbitrary function. From (mt2.2) (the 'easier' of the two to integrate)

$$F(x,y) = x^3e^y + f(y) \qquad \text{(mt2.4)}$$

where f is an arbitrary function of y. The F given by (mt2.4) must also satisfy (mt2.3) - this will determine $f(y)$. From (mt2.4)

$$\frac{\partial F}{\partial y} = x^3e^y + f'(y)$$

Comparing this with (mt2.3) we see that $f'(y) = \sin y$ so that $f(y) = -\cos y + c_2$. From (mt2.4), we have that the function F is given by

$$F(x,y) = x^3e^y - \cos y + c_2$$

Finally, the general solution of the differential equation is given by

$$\begin{aligned} F(x,y) &= c_1 \\ x^3e^y - \cos y + c_2 &= c_1 \\ x^3e^y - \cos y &= c_3, \qquad c_3 = c_1 - c_2 \end{aligned}$$

Question 4.(ii) The equation

$$\cos y\frac{dy}{dx} = xe^x$$

is *separable* (see Section 2.1). Hence,

$$\begin{aligned} \int \cos y dy &= \int xe^x dx \\ \sin y &= xe^x - \int 1 \cdot e^x dx \qquad \text{(integration by parts - Appendix)} \\ &= xe^x - e^x + c_1 \\ &= (x-1)\, e^x + c_1 \end{aligned}$$

Finally, the general solution is given by

$$\sin y = [(x-1)\, e^x + c_1]$$

Question 4.(iii) The equation

$$\frac{dy}{dx} = \frac{y+x}{x}$$

can be considered either as a *homogeneous* equation (Section 2.2) or one which is *linear in y* (Section 2.5). Taking the latter approach

$$\frac{dy}{dx} - \frac{y}{x} = 1$$

An integrating factor is given by

$$\begin{aligned} \mu(x) &= \exp\left(\int -\frac{1}{x}dx\right) \\ &= \exp\left(-\ln|x|\right) \\ &= \frac{1}{|x|} \end{aligned}$$

Choose

$$\mu(x) = \frac{1}{x}$$

Hence, the differential equation can be written as

$$\frac{d}{dx}\left(\frac{y}{x}\right) = \frac{1}{x}$$

Integrate both sides with respect to x.

$$\begin{aligned} \frac{y}{x} &= \int \frac{1}{x} dx = \ln|x| + c_1 \\ &= \ln A|x|, \qquad c_1 = \ln A, \quad A > 0 \end{aligned}$$

Hence, the general solution is given by

$$y(x) = x \ln A|x|$$

Question 5. (15%) Solve the following initial value problem

$$\frac{dy}{dx} = \frac{y}{y^2 e^y + x}, \qquad y(0) = 1$$

Solution. Writing the equation in the form

$$\begin{aligned} \frac{dx}{dy} &= \frac{y^2 e^y + x}{y} \\ &= y e^y + \frac{x}{y} \end{aligned}$$

or

$$\frac{dx}{dy} - \frac{x}{y} = y e^y$$

we see that the equation is, in fact, *linear in* x (see Note 2.5.2). An integrating factor is given by

$$\mu(y) = \exp\left(-\int \frac{1}{y} dy\right) = \frac{1}{|y|}$$

Choose,

$$\mu(y) = \frac{1}{y}$$

The differential equation can now be written as

$$\frac{d}{dy}\left(\frac{x}{y}\right) = \frac{y e^y}{y} = e^y$$

Integrating both sides with respect to y, we obtain

$$\begin{aligned} \frac{x}{y} &= e^y + c_1 \\ x(y) &= y e^y + y c_1 \end{aligned}$$

Now, the given initial condition states that $y(0) = 1$ or $x = 0$ when $y = 1$. Substituting this information into the general solution

$$x = ye^y + yc_1$$

we obtain

$$\begin{aligned} 0 &= 1 \cdot e + c_1 \\ c_1 &= -e \end{aligned}$$

Hence, the solution of the intial value problem is given by

$$\begin{aligned} x &= ye^y + y(-e) \\ &= y(e^y - e) \end{aligned}$$

MIDTERM EXAMINATION #3
SOLUTIONS

In what follows, c_i, $i = 1, 2, \dots$ will denote arbitrary constants. Rules of logarithms and exponentials, a summary of the main techniques of integration as well as a *table of integrals* can be found in the Appendix.

Question 1.(20%) Consider the pair of differential equations

$$\begin{aligned}(x-1)\,y'' - xy' + y &= 0 \qquad \text{(mt3.1)}\\ (x-1)\,y'' - xy' + y &= 1\end{aligned}$$

Given that $y_1(x) = e^x$ solves the homogeneous differential equation (mt3.1), find the general solution of each differential equation.

Solution. The fact that we are given one solution of the homogeneous equation (mt3.1) suggests that we can use *reduction of order* (see Section 3.4) to solve the *inhomogeneous* equation. Since the latter is linear, its general solution will decompose into two parts: the part attached to arbitrary constants will be the complementary solution (see Section 3.1), that is, the general solution of the *homogeneous* equation (mt3.1). Hence, the general solution of the equation

$$(x-1)\,y'' - xy' + y = 1 \qquad \text{(mt3.2)}$$

is sought in the form $y = y_1(x)v(x) = ve^x$ where v is a function of x, to be determined from some first order differential equation.

$$\begin{aligned} y &= ve^x \\ y' &= ve^x + v'e^x \\ &= e^x\,(v + v') \\ y'' &= e^x\,(v+v') + e^x\,(v' + v'') \\ &= e^x\,(v'' + 2v' + v) \end{aligned}$$

Substituting these expressions into the differential equation (mt3.2), we obtain

$$\begin{aligned} (x-1)\,e^x\,(v'' + 2v' + v) - xe^x\,(v + v') + ve^x &= 1 \\ e^x\,(x-1)\,v'' + e^x\,(x-2)\,v' &= 1 \end{aligned}$$

Let $w = v'$.

$$w' + \left(\frac{x-2}{x-1}\right) w = \frac{1}{e^x (x-1)} \tag{mt3.3}$$

This equation is linear in w (Section 2.5). An integrating factor is given by

$$\mu(x) = \exp\left(\int \frac{x-2}{x-1} dx\right)$$

The integrand can be written as

$$\frac{x-2}{x-1} = 1 - \frac{1}{x-1}$$

Hence,

$$\begin{aligned} \mu(x) &= \exp\left(\int \left[1 - \frac{1}{x-1}\right] dx\right) \\ &= e^x e^{-\ln|x-1|} \\ &= \frac{e^x}{|x-1|} \end{aligned}$$

Choose,

$$\mu(x) = \frac{e^x}{x-1}$$

The differential equation (mt3.3) now becomes

$$\frac{d}{dx}\left(w\left[\frac{e^x}{x-1}\right]\right) = \left(\frac{e^x}{x-1}\right)\left(\frac{1}{e^x (x-1)}\right)$$

Integrating both sides with respect to x, we obtain

$$\begin{aligned} w\left(\frac{e^x}{x-1}\right) &= \int \frac{1}{(x-1)^2} dx \\ &= -(x-1)^{-1} + c_1 \end{aligned}$$

Thus,

$$\begin{aligned} w(x) &= \left[-(x-1)^{-1} + c_1\right] \frac{x-1}{e^x} \\ &= -e^{-x} + c_1 (x-1) e^{-x} \end{aligned}$$

Hence,

$$\begin{aligned} v(x) &= \int w(x) dx \\ &= \int \left[-e^{-x} + c_1 (x-1) e^{-x}\right] dx \\ &= e^{-x} + c_1 e^{-x} + c_1 \underbrace{\int x e^{-x} dx}_{\text{Int. by Parts}} \\ &= e^{-x} + c_1 e^{-x} + c_1 \left(-x e^{-x} - e^{-x}\right) + c_2 \\ &= e^{-x} - c_1 x e^{-x} + c_2 \end{aligned}$$

Finally, the general solution of the inhomogeneous equation (mt3.2) is given by

$$\begin{aligned} y(x) &= v(x)e^x \\ &= \left(e^{-x} - c_1 x e^{-x} + c_2\right) e^x \\ &= 1 - c_1 x + c_2 e^x \end{aligned}$$

Since the equation is linear, we can identify

$$\begin{aligned} y_p(x) &= 1 \\ y_c(x) &= -c_1 x + c_2 e^x \\ &= c_3 x + c_2 e^x, \quad c_3 = -c_1 \end{aligned}$$

Hence, the general solution of the homogeneous equation (mt3.1) is given by

$$y(x) = c_3 x + c_2 e^x$$

2. Find the general solution

(i) (10%)

$$y(x^2+2)dx = (x^3 - x)dy$$

(ii) (10%)

$$\frac{dy}{dx} = \frac{x^4 \cos 3x + 3y}{x}$$

Solution.

Question 2.(i) Writing the equation in the form

$$\frac{dy}{dx} = \frac{y\left(x^2+2\right)}{x^3 - x} = y \cdot \frac{\left(x^2+2\right)}{x^3 - x}$$

we see that the equation is separable (see Section 2.1). Consequently,

$$\int \frac{dy}{y} = \int \frac{x^2+2}{x^3-x} dx$$

Using partial fractions (Appendix) for the integral on the right-hand side, we have

$$\begin{aligned} \ln|y| &= \int \left[-\frac{2}{x} + \frac{3}{2(x-1)} + \frac{3}{2(x+1)}\right] dx \\ &= -2\ln|x| + \frac{3}{2}\ln|x-1| + \frac{3}{2}\ln|x+1| + c_1 \end{aligned}$$

or

$$2\ln|y| = -4\ln|x| + 3\ln|x-1| + 3\ln|x+1| + 2c_1$$

Combining the logarithms (Appendix) and setting $2c_1 = \ln A,\ \ A > 0$,, we have

$$\ln|y|^2 = \ln A\,|x|^{-4}\,|x-1|^3\,|x+1|^3$$

Taking exponentials of both sides

$$y^2 = A\left|\frac{(x-1)^3(x+1)^3}{x^4}\right|$$

Setting $B = \pm A$, the general solution is given by

$$y^2 = B\frac{(x-1)^3(x+1)^3}{x^4}$$

Question 2.(ii) The differential equation

$$\frac{dy}{dx} = \frac{x^4\cos 3x + 3y}{x}$$

is *linear in y* (Section 2.5). Writing the equation in standard form, we have

$$\frac{dy}{dx} - \frac{3y}{x} = x^3\cos 3x \qquad \text{(mt3.4)}$$

so that an integrating factor is given by

$$\begin{aligned}\mu(x) &= \exp\left(-\int\frac{3}{x}dx\right)\\ &= \exp(-3\ln|x|)\\ &= \frac{1}{|x|^3}\end{aligned}$$

Choose,

$$\mu(x) = \frac{1}{x^3}$$

The differential equation (mt3.4) can now be written as

$$\frac{d}{dx}\left(\frac{y}{x^3}\right) = \cos 3x$$

Integrating both sides with respect to x, we obtain

$$\frac{y}{x^3} = \frac{1}{3}\sin 3x + c_1$$

The general solution is therefore given by

$$y(x) = \frac{x^3}{3}\sin 3x + x^3 c_1$$

Question 3.(15%) Find the general solution of the equation

$$\frac{d^2y}{dx^2} - 4\frac{dy}{dx} + 4y = 6xe^{2x}$$

Solution. Since the differential equation is linear, the general solution $y(x)$ decomposes into two parts (see Section 3.1):

$$y(x) = y_c(x) + y_p(x)$$

For $y_c(x)$, the characteristic equation is given by

$$\begin{aligned} m^2 - 4m + 4 &= 0 \\ (m-2)^2 &= 0 \end{aligned}$$

Hence, from Table 3.2.1,

$$y_c(x) = (c_1 + c_2 x)\, e^{2x}$$

For $y_p(x)$, using Table 3.3.1, try

$$y_p(x) = \underset{\substack{\uparrow \\ (*)}}{x^2}\,(A + Bx)\, e^{2x}$$

Note that the multiplicative term x^2 at (*) is necessary to avoid any terms in $y_p(x)$ common to both y_c and y_p. To find the definite constants A and B , we substitute $y_p(x)$ into the differential equation.

$$\begin{aligned} y_p(x) &= x^2(A+Bx)\,e^{2x} \\ y_p'(x) &= 2e^{2x}\left(Ax^2 + Bx^3\right) + e^{2x}\left(2Ax + 3Bx^2\right) \\ &= e^{2x}[2Bx^3 + x^2(2A+3B) + 2Ax] \\ y_p''(x) &= 2e^{2x}[2Bx^3 + x^2(2A+3B) + 2Ax] + e^{2x}[6Bx^2 + 2x(2A+3B) + 2A] \\ &= e^{2x}[4Bx^3 + (4A+12B)x^2 + (8A+6B)x + 2A] \end{aligned}$$

The differential equation becomes

$$\begin{aligned} \frac{d^2y_p}{dx^2} - 4\frac{dy_p}{dx} + 4y_p &= 6xe^{2x} \\ (6Bx + 2A)\,e^{2x} &= 6xe^{2x} \end{aligned}$$

Hence,

$$B = 1 \text{ and } A = 0$$

so that

$$y_p(x) = x^3 e^{2x}$$

Finally, the general solution is given by

$$\begin{aligned} y(x) &= y_c(x) + y_p(x) \\ &= (c_1 + c_2 x) e^{2x} + x^3 e^{2x} \end{aligned}$$

Question 4.(15%) Solve the equation

$$\frac{dy}{dx} = \frac{1 - xy}{x(x - y)}$$

Solution. The differential equation is neither separable, exact, homogeneous, linear nor Bernouilli. We try to find an integrating factor (see Section 2.4).

We first write the differential equation in the standard form (2.3.1).

$$\underbrace{(xy - 1)}_{M} dx + \underbrace{\left(x^2 - xy\right)}_{N} dy = 0$$

Next, we determine the required partial derivatives with each of (2.4.1) and (2.4.3) in mind.

$$\frac{\partial M}{\partial y} = x; \qquad \frac{\partial N}{\partial x} = 2x - y$$

Examine (2.4.1)

$$\begin{aligned} \frac{1}{N}\left(\frac{\partial M}{\partial y} - \frac{\partial N}{\partial x}\right) &= \frac{1}{x^2 - xy}(x - 2x + y) \\ &= \frac{1}{x(x - y)}(y - x) \\ &= -\frac{1}{x} \\ &= R(x) \end{aligned}$$

Clearly, (2.4.1) is a function of x only!
Hence, an integrating factor $\mu(x)$ can be found from (2.4.2).

$$\begin{aligned} \mu(x) &= \exp(-\int \frac{1}{x} dx) \\ &= \exp(-\ln|x|) \\ &= |x|^{-1} + c_1 \end{aligned}$$

Choose

$$\mu(x) = x^{-1}$$

Next, we multiply both sides of our differential equation by $\mu(x) = x^{-1}$.

$$\underbrace{x^{-1}(xy-1)}_{M}\,dx + \underbrace{x^{-1}(x^2-xy)}_{N}\,dy = 0$$

This equation is now exact (in the sense of Section 2.3) since

$$\frac{\partial M}{\partial y} = 1 = \frac{\partial N}{\partial x}$$

Its general solution can be now be found as in Section 2.3 as follows. The resulting exact equation is given by

$$\underbrace{\left(y - \frac{1}{x}\right)}_{M} dx + \underbrace{(x-y)}_{N}\,dy = 0$$

Hence, there exists a function $F(x,y)$ such that

$$\frac{\partial F}{\partial x} = M = y - \frac{1}{x} \qquad \text{(mt3.5)}$$

$$\frac{\partial F}{\partial y} = N = x - y \qquad \text{(mt3.6)}$$

From (mt3.6)

$$F(x,y) = xy - \frac{y^2}{2} + f(x) \qquad \text{(mt3.7)}$$

where f is arbitrary. From (mt3.5) and(mt3.6), we require

$$y + f'(x) = y - \frac{1}{x}$$

It follows that

$$\begin{aligned} f'(x) &= -\frac{1}{x} \\ f(x) &= -\ln|x| + c_2 \end{aligned}$$

Finally, from (mt3.7), the general solution is given by

$$\begin{aligned} F(x,y) &= c_3 \\ xy - \frac{y^2}{2} - \ln|x| + c_2 &= c_3 \\ xy - \frac{y^2}{2} - \ln|x| &= c_4 \end{aligned}$$

where $c_4 = c_3 - c_2$.

Question 5.(20%) Find the general solution of

$$\frac{d^2y}{dx^2} + 2\frac{dy}{dx} + y = \frac{1}{(e^x - 1)^2}$$

Solution. The right-hand side of the differential equation is 'non-standard' in that it doesn't appear in Table 3.3.1. Hence, we cannot apply the method of undetermined coefficients. However, the differential equation is linear with constant coefficients. Thus we can find y_c and use one of these solutions in a *reduction of order* process (see Section 3.4) to find y_p. In fact, the general solution of the equation is given by (Section 3.1)

$$y(x) = y_c(x) + y_p(x)$$

For $y_c(x)$, the characteristic equation is given by

$$\begin{aligned} m^2 + 2m + 1 &= 0 \\ (m+1)^2 &= 0 \end{aligned}$$

Hence, from Table 3.2.1,

$$y_c(x) = (c_1 + c_2 x) e^{-x}$$

For $y_p(x)$, using reduction of order, we choose the solution $y_1(x) = e^{-x}$ from $y_c(x)$ (remember, we need only *one* solution of the homogeneous equation for a reduction of order process) and suggest

$$y_p(x) = ve^{-x}$$

where v is a function of x to be determined. Substitute this y_p into the differential equation. In fact,

$$\begin{aligned} y_p'(x) &= (v' - v) e^{-x} \\ y_p''(x) &= e^{-x}(v'' - 2v' + v) \end{aligned}$$

so that

$$\begin{aligned} \frac{d^2y_p}{dx^2} + 2\frac{dy_p}{dx} + y_p &= \frac{1}{(e^x - 1)^2} \\ [(v'' - 2v' + v) + 2(v' - v) + v] e^{-x} &= \frac{1}{(e^x - 1)^2} \\ v'' e^{-x} &= \frac{1}{(e^x - 1)^2} \\ v'' &= \frac{e^x}{(e^x - 1)^2} \end{aligned}$$

Hence,

$$v' = \int \frac{e^x}{(e^x-1)^2} dx$$

To find this integral, let $u = e^x - 1$, $du = e^x dx = (u+1)\, dx$

$$\begin{aligned} v'(x) &= \int \frac{1}{u^2} du \\ &= -u^{-1} + c_3 \\ &= -(e^x - 1)^{-1} + c_3 \end{aligned}$$

Since we are interested only in a particular solution y_p, we set all constants of integration to zero here and in future integrations. Thus

$$v'(x) = -(e^x - 1)^{-1}$$

Next,

$$v(x) = -\int \frac{dx}{e^x - 1}$$

Again, to find this integral, let $u = e^x - 1$.

$$\begin{aligned} v(x) &= -\int \frac{du}{u(u+1)} \\ &= -\int \left(\frac{1}{u} - \frac{1}{u+1}\right) du \\ &= -(\ln|e^x - 1| - \ln e^x) \\ &= -\ln|e^x - 1| + x \end{aligned}$$

Finally,

$$\begin{aligned} y_p(x) &= e^{-x} v(x) \\ &= e^{-x}(x - \ln|e^x - 1|) \end{aligned}$$

The general solution is now given by

$$\begin{aligned} y(x) &= y_c(x) + y_p(x) \\ &= (c_1 + c_2 x) e^{-x} + x e^{-x} - e^{-x} \ln|e^x - 1| \\ &= e^{-x}[c_1 + c_3 x - \ln|e^x - 1|] \end{aligned}$$

where, $c_3 = c_2 + 1$.

Note. Since we know y_c entirely, we can also find y_p using *variation of parameters* (Section 3.5). In fact, for $y_p(x)$, using the variation of parameters technique, from the form of $y_c(x)$, the particular solution is given by

$$y_p(x) = A(x) e^{-x} + x B(x) e^{-x}$$

where A and B are functions determined from the system

$$\begin{bmatrix} e^{-x} & xe^{-x} \\ -e^{-x} & e^{-x}(1-x) \end{bmatrix} \begin{bmatrix} A' \\ B' \end{bmatrix} = \begin{bmatrix} 0 \\ (e^x - 1)^{-2} \end{bmatrix}$$

Solving this matrix equation (or using Cramer's rule to solve the equivalent system), we obtain

$$\begin{bmatrix} A' \\ B' \end{bmatrix} = \begin{bmatrix} -xe^x (e^x - 1)^{-2} \\ e^x (e^x - 1)^{-2} \end{bmatrix}$$

Hence,

$$\begin{aligned} A(x) &= -\int xe^x (e^x - 1)^{-2} \, dx \\ B(x) &= \int e^x (e^x - 1)^{-2} \, dx \end{aligned}$$

To find $B(x)$, let $u = e^x - 1$, $du = e^x dx = (u+1)\, dx$

$$\begin{aligned} B(x) &= \int \frac{1}{u^2} du \\ &= -u^{-1} + c_3 \\ &= -(e^x - 1)^{-1} + c_3 \end{aligned}$$

Since we are interested only in a particular solution, we set all constants of integration to zero. To find $A(x)$, use integration by parts (Appendix).

$$\begin{aligned} A(x) &= -\left[xB(x) - \int B(x) dx \right] \\ &= -xB(x) - \underbrace{\int (e^x - 1)^{-1} \, dx}_{\text{Let } u = e^x - 1} \\ &= -xB(x) - \int \frac{1}{u(u+1)} du \\ &= -xB(x) - \left[\int \left(\frac{1}{u} - \frac{1}{u+1} \right) du \right] \\ &= -xB(x) - (\ln |e^x - 1| - \ln e^x) \\ &= \frac{x}{e^x - 1} - \ln |e^x - 1| + x \end{aligned}$$

Thus,

$$\begin{aligned} y_p(x) &= A(x) e^{-x} + xB(x) e^{-x} \\ &= \left(\frac{x}{e^x - 1} - \ln |e^x - 1| + x \right) e^{-x} - xe^{-x} \left(\frac{1}{e^x - 1} \right) \\ &= xe^{-x} - e^{-x} \ln |e^x - 1| \end{aligned}$$

Finally, the general solution is given by

$$\begin{aligned} y(x) &= y_c(x) + y_p(x) \\ &= (c_1 + c_2 x) e^{-x} + x e^{-x} - e^{-x} \ln |e^x - 1| \\ &= e^{-x} [c_1 + c_3 x - \ln |e^x - 1|] \end{aligned}$$

where, $c_3 = c_2 + 1$ - as before!

Question 6.(10%) The current i at time t in an electrical circuit is modelled by the differential equation

$$\frac{di}{dt} = \frac{V - Ri}{L}$$

where V is voltage, R is resistance and L is inductance. If V, R and L are positive constants and the current is initially at zero, show that the current (eventually) reaches a steady state of $i = i_c = \frac{V}{R}$.

Solution. Writing the differential equation in the form

$$\frac{di}{dt} + \frac{Ri}{L} = \frac{V}{L}$$

we see that it is, in fact, *linear* in i (see Section 2.5). An integrating factor is given by

$$\begin{aligned} \mu(t) &= \exp\left(\int \frac{R}{L} dt\right) \\ &= \exp\left(\frac{R}{L} t\right) \end{aligned}$$

Thus, the equation can be written in the form

$$\frac{d}{dt}\left(i e^{\frac{R}{L}t}\right) = \frac{V}{L} e^{\frac{R}{L}t}$$

Integrating both sides with respect to t, we have

$$\begin{aligned} i e^{\frac{R}{L}t} &= \frac{V}{L} \int e^{\frac{R}{L}t} dt \\ &= \frac{V}{L} \cdot \frac{L}{R} e^{\frac{R}{L}t} + c_1 \\ &= \frac{V}{R} e^{\frac{R}{L}t} + c_1 \end{aligned}$$

Hence, the general solution is given by

$$i(t) = \frac{V}{R} + c_1 e^{-\frac{R}{L}t} \tag{mt3.8}$$

Initially, the current is zero so that $i(0) = 0$. This enables us to find the constant c_1. In fact, from (mt3.8),

$$\begin{aligned} 0 &= \frac{V}{R} + c_1 \\ c_1 &= -\frac{V}{R} \end{aligned}$$

Thus, the particular solution is given by

$$i(t) = \frac{V}{R}\left(1 - e^{-\frac{R}{L}t}\right)$$

It is clear that since $R, L > 0$, as $t \longrightarrow \infty$, $e^{-\frac{R}{L}t} \longrightarrow 0$ and $i(t) \longrightarrow \frac{V}{R}$. Thus, after a sufficient amount of time has elapsed, the current does indeed reach a steady state of

$$i = i_c = \frac{V}{R}$$

MIDTERM EXAMINATION #4

SOLUTIONS

In what follows, c_i, $i = 1, 2, ...$ will denote arbitrary constants. Rules of logarithms and exponentials, a summary of the main techniques of integration as well as a *table of integrals* can be found in the Appendix.

Question 1.(15%) Show that the solution of the initial value problem

$$\left[t\cos^2\left(\frac{y}{t}\right) - y\right] dt = -tdy, \qquad y(1) = \frac{\pi}{4}$$

is given by $\tan\dfrac{y}{t} = \ln\dfrac{e}{|t|}$.

Solution. Try to identify the differential equation by writing it in standard form.

$$\frac{dy}{dt} = \frac{y - t\cos^2\left(\frac{y}{t}\right)}{t}$$

The equation is not separable but may be homogeneous (see Section 2.2). To confirm homogeneity, let

$$f(t,y) = \frac{y - t\cos^2\left(\frac{y}{t}\right)}{t}$$

Then

$$\begin{aligned} f(\lambda t, \lambda y) &= \frac{\lambda y - \lambda t\cos^2\left(\frac{\lambda y}{\lambda t}\right)}{\lambda t} \\ &= \frac{y - t\cos^2\left(\frac{y}{t}\right)}{t} \\ &= f(t,y) \end{aligned}$$

Hence, the equation is indeed homogeneous. Let $y = vt$ where v is a function of t to be determined. Substituting into the differential equation, we obtain

$$\begin{aligned} v + t\frac{dv}{dt} &= \frac{vt - t\cos^2\left(\frac{vt}{t}\right)}{t} \\ &= v - \cos^2 v \end{aligned}$$

Hence, we obtain the separable equation

$$t\frac{dv}{dt} = -\cos^2 v$$

$$
\begin{aligned}
-\int \frac{dv}{\cos^2 v} &= \int \frac{dt}{t} \\
-\int \sec^2 v dv &= \ln|t| + c_1 \\
-\tan v &= \ln|t| + c_1
\end{aligned}
$$

Next, let $v = \frac{y}{t}$ to obtain the general solution

$$
\tan \frac{y}{t} = -\ln|t| + c_1
$$

From the given initial condition, $y(1) = \frac{\pi}{4}$, we have that $y = \frac{\pi}{4}$ when $t = 1$ so that

$$
\begin{aligned}
\tan \frac{\pi}{4} &= -\ln 1 + c_1 \\
1 &= c_1
\end{aligned}
$$

Finally, the solution of the initial value problem is given by

$$
\begin{aligned}
\tan \frac{y}{t} &= -\ln|t| + 1 \\
&= -\ln|t| + \ln e \\
&= \ln e - \ln|t|
\end{aligned}
$$

or

$$
\tan \frac{y}{t} = \ln \frac{e}{|t|}
$$

Question 2.(20%) Find the orthogonal trajectories corresponding to the family of curves with parameter k described by the equation $y^2 = x^2(1 - kx)$. In particular, find the orthogonal trajectory passing through the point $(1, 1)$.

Solution. According to Section 2.8, we first find the differential equation corresponding to the family of curves given by

$$
y^2 = x^2(1 - kx)
$$

To do this, we differentiate implicitly with respect to x.

$$
\begin{aligned}
2yy' &= 2x - 3kx^2 \\
y' &= \frac{x(2 - 3kx)}{2y}
\end{aligned}
$$

However, since $y^2 = x^2(1 - kx)$, we have

$$
k = \frac{x^2 - y^2}{x^3}
$$

Hence, the differential equation for the family of curves becomes

$$\begin{aligned} y' &= \frac{x\left(2 - 3x\left[\frac{x^2-y^2}{x^3}\right]\right)}{2y} \\ &= \frac{3y^2 - x^2}{2xy} \end{aligned}$$

From Section 2.8, the differential equation of the family of *orthogonal trajectories* is given by

$$\frac{dy}{dx} = \frac{2xy}{x^2 - 3y^2}$$

This equation is *homogeneous* (see Section 2.2) (to check this, merely let $f(x,y) = \dfrac{2xy}{x^2 - 3y^2}$ and note that $f(x,y) = f(\lambda x, \lambda y)$). If we write the equation in the form

$$2xydx + (3y^2 - x^2)dy = 0$$

from Note 2.2.2, we see that it may be simpler to substitute for $x = vy$. The equation then becomes

$$\begin{aligned} 2vy^2\left(vdy + ydv\right) + \left(3y^2 - v^2y^2\right)dy &= 0 \\ 2vydv + (v^2 + 3)dy &= 0 \end{aligned}$$

or

$$\frac{dy}{y} = -\frac{2vdv}{v^2 + 3}$$

As expected, the equation is now *separable.* Following the procedure in Section 2.1, we have

$$\begin{aligned} \int \frac{dy}{y} &= -\int \frac{2vdv}{v^2 + 3} \\ \ln|y| &= -\ln\left(v^2 + 3\right) + c_1 \\ &= -\ln A\left(v^2 + 3\right) \\ &= \ln[A\left(v^2 + 3\right)]^{-1} \end{aligned}$$

where $c_1 = -\ln A,\ A > 0$. Taking exponentials of both sides, we have

$$\begin{aligned} |y| &= \frac{1}{A(v^2 + 3)} \\ y &= \frac{B}{v^2 + 3} \end{aligned}$$

where $B = \pm\dfrac{1}{A}$. Finally, let $v = \dfrac{x}{y}$.

$$y = \frac{B}{\left(\frac{x}{y}\right)^2 + 3}$$

so that the equation of the family of orthogonal trajectories is given by

$$By = x^2 + 3y^2$$

To obtain the particular trajectory passing through the point $(1,1)$, we let $x = y = 1$:

$$B = 1 + 3 = 4$$

Hence, the particular orthogonal trajectory passing through the point $(1,1)$ is given by

$$4y = x^2 + 3y^2$$

Note. The homogeneous equation above can also be solved using the substitution $y = vx$. The only difference is that the resulting integrations are slightly more complicated.

3. Find the general solution.

(i) (10%)

$$D^2(D^2 - 2D + 2)^2(D^2 - 3D + 2)y = f(x) \qquad \text{(mt4.1)}$$

where $f(x) = 0$, $D^n y \equiv \dfrac{d^n y}{dx^n}$, $n = 1, 2, \ldots$ and y is a function of the independent variable x.

(ii) (10%) $\cos\theta \dfrac{dr}{d\theta} = -\sin\theta\,(1 - 2r)$, where r is a function of θ.

(iii) (15%) $\dfrac{d^2y}{dx^2} + y = \sec x$

Solution

Question 3.(i) The equation

$$D^2(D^2 - 2D + 2)^2(D^2 - 3D + 2)y = 0$$

is linear, homogeneous with constant coefficients (see Section 3.2). To find its general solution, first find the corresponding characteristic equation.

$$\begin{aligned} m^2\left(m^2 - 2m + 2\right)^2\left(m^2 - 3m + 2\right) &= 0 \\ m^2\left(m^2 - 2m + 2\right)^2(m - 2)(m - 1) &= 0 \end{aligned}$$

The solutions are

$$\begin{aligned} m &= 0 \ \text{(twice)} \\ m &= 1 \pm i \ \text{(twice)} \\ m &= 1, 2 \end{aligned}$$

From Table 3.2.1, the general solution is given by

$$y(x) = \underbrace{c_1 + c_2 x}_{m=0} + \underbrace{e^x\left[(c_3 + c_4 x)\cos x + (c_5 + c_6 x)\sin x\right]}_{m=1\pm i} + \underbrace{c_7 e^x}_{m=1} + \underbrace{c_8 e^{2x}}_{m=2}$$

Question 3.(ii) Write the equation in standard form.

$$\begin{aligned}\cos\theta\frac{dr}{d\theta} &= -\sin\theta - 2r\sin\theta\\ \frac{dr}{d\theta} + 2r\tan\theta &= -\tan\theta\end{aligned}$$

The equation is *linear in r* (see Section 2.5). An integrating factor is given by

$$\begin{aligned}\mu(\theta) &= \exp\left(\int 2\tan\theta d\theta\right)\\ &= \exp\left(-2\ln|\cos\theta|\right)\\ &= \sec^2\theta\end{aligned}$$

The differential equation can now be written in the form

$$\frac{d}{d\theta}\left(r\sec^2\theta\right) = -\sec^2\theta\tan\theta$$

Integrate both sides with respect to θ to obtain the general solution.

$$\begin{aligned}r\sec^2\theta &= -\underbrace{\int \sec^2\theta\tan\theta d\theta}_{\text{Let } u=\tan\theta}\\ r\sec^2\theta &= -\frac{\tan^2\theta}{2} + c_1\\ r &= -\frac{\tan^2\theta}{2\sec^2\theta} + c_1\cos^2\theta\\ r &= -\frac{1}{2}\sin^2\theta + c_1\cos^2\theta\end{aligned}$$

(Note that this can also be written in the form $r = -\frac{1}{2} + c_2\cos^2\theta,\ \ c_2 = c_1 + \frac{1}{2}$)

Question 3.(iii) The differential equation

$$\frac{d^2y}{dx^2} + y = \sec x$$

is linear with constants coefficients but the right-hand side is not in a form suitable for the method of undetermined coefficients. However, we can easily find y_c which will provide us with sufficient information for a *reduction of order* (Section 3.4) solution.

$$y(x) = y_c(x) + y_p(x)$$

For $y_c(x)$, the characteristic equation is given by

$$\begin{aligned} m^2 + 1 &= 0 \\ m &= \pm i \end{aligned}$$

From Table 3.2.1,

$$y_c(x) = c_1 \cos x + c_2 \sin x$$

For $y_p(x)$, we use reduction of order. Let

$$\begin{aligned} y_p(x) &= v \cos x \\ y_p'(x) &= v' \cos x - v \sin x \\ y_p''(x) &= v'' \cos x - v' \sin x - v' \sin x - v \cos x \\ &= v'' \cos x - 2v' \sin x - v \cos x \end{aligned}$$

where v is to be determined. Then the differential equation becomes

$$\begin{aligned} v'' \cos x - 2v' \sin x - v \cos x + v \cos x &= \sec x \\ v'' \cos x - 2v' \sin x &= \sec x \end{aligned}$$

Let $w = v'$.

$$w' - 2w \tan x = \sec^2 x$$

This equation is linear in w (see Section 2.5). An integrating factor is given by

$$\begin{aligned} \mu(x) &= \exp\left(-\int 2 \tan x dx\right) \\ &= \cos^2 x \end{aligned}$$

The differential equation in w now becomes

$$\frac{d}{dx}\left(w \cos^2 x\right) = \sec^2 x \cos^2 x = 1$$

Integrate both sides with respect to x - since we are interested only in a particular solution y_p, we set all constants of integration to zero..

$$\begin{aligned} w \cos^2 x &= x \\ w &= x \sec^2 x \end{aligned}$$

Hence,

$$\begin{aligned} v &= \int w dx \\ &= \underbrace{\int x \sec^2 x dx}_{\text{by Parts}} \\ &= x \tan x - \int \tan x dx \\ &= x \tan x + \ln |\cos x| \end{aligned}$$

Finally,

$$\begin{aligned} y_p(x) &= v \cos x \\ &= x \sin x + \cos x \ln |\cos x| \end{aligned}$$

Hence, the general solution is given by

$$\begin{aligned} y(x) &= y_c(x) + y_p(x) \\ &= c_1 \cos x + c_2 \sin x + x \sin x + \cos x \ln |\cos x| \end{aligned}$$

Question 4.(10%) Consider the above differential equation (mt4.1). Suppose the right-hand side is replaced by the function $f(x) = 3 + xe^x \sin x + 3e^{4x}$. If a particular solution $y_p(x)$ is sought using the method of undetermined coefficients, write down the *form* of $y_p(x)$ **- do not evaluate the coefficients.**

Solution. For the linear equation

$$D^2(D^2 - 2D + 2)^2(D^2 - 3D + 2)y = 3 + xe^x \sin x + 3e^{4x}$$

we know that (Section 3.1)

$$y(x) = y_c(x) + y_p(x)$$

From Q3(i), we have that

$$y_c(x) = c_1 + c_2 x + e^x [(c_3 + c_4 x) \cos x + (c_5 + c_6 x) \sin x] + c_7 e^x + c_8 e^{2x}$$

Since the right-hand side of the differential equation has three terms, using Table 3.3.1, we have the following form for $y_p(x)$.

$$\begin{aligned} y_p(x) &= y_{p_1}(x) + y_{p_2}(x) + y_{p_3}(x) \\ y_{p_1}(x) &= \underset{\substack{\uparrow \\ (*)}}{x^2} A \\ y_{p_2}(x) &= \underset{\substack{\uparrow \\ (*)}}{x^2} e^x [(B + Cx) \cos x + (E + Fx) \sin x] \\ y_{p_3}(x) &= Ge^{4x} \end{aligned}$$

where A, B, C, E, F and G are definite constants to be determined. Note that in each case, at (*), it is necessary to multiply the original suggestion for the corresponding term in $y_p(x)$ by x^2 to avoid any terms in common with $y_c(x)$.

Question 5. (10%) Solve

$$\frac{dy}{dx} = \frac{-y\cos x - \cos y}{\sin x - x\sin y}$$

given that $y(1) = 0$.

Solution. Writing the equation in the form

$$\underbrace{(y\cos x + \cos y)}_{M(x,y)}\, dx + \underbrace{(\sin x - x\sin y)}_{N(x,y)}\, dy = 0$$

we note that

$$\frac{\partial M}{\partial y} = \cos x - \sin y = \frac{\partial N}{\partial x}$$

so that the differential equation is exact (see Section 2.3). Hence, there exists a function $F(x, y)$ such that

$$\frac{\partial F}{\partial x} = M(x,y) = y\cos x + \cos y \qquad \text{(mt4.2)}$$

$$\frac{\partial F}{\partial y} = N(x,y) = \sin x - x\sin y \qquad \text{(mt4.3)}$$

and the general solution of the equation is given by

$$F(x,y) = c_1 \qquad \text{(mt4.4)}$$

From (mt4.2), we have

$$F(x,y) = y\sin x + x\cos y + f(y) \qquad \text{(mt4.5)}$$

where f is an arbitrary function of y. The F from (mt4.5) must also satisfy (mt4.3). Hence, we must have

$$\sin x - x\sin y + f'(y) = \sin x - x\sin y$$

Hence, we must have that $f'(y) = 0$ or $f(y) = c_2$. Finally, from (mt4.5), we have that

$$F(x,y) = y\sin x + x\cos y + c_2$$

so that, from (mt4.4), the general solution of the equation is given by

$$\begin{aligned} F(x,y) &= c_1 \\ y\sin x + x\cos y + c_2 &= c_1 \\ y\sin x + x\cos y &= c_3 \end{aligned}$$

where $c_3 = c_1 - c_2$.

Question 6.(10%) Show that the functions

$$y_1(x) = x, \quad y_2(x) = \frac{1}{x}, \quad y_3(x) = x^2$$

are linearly independent solutions of the equation

$$x^3 y''' + x^2 y'' - 2xy' + 2y = 0, \quad x \neq 0$$

Hence or otherwise, find the general solution.

Solution. Substituting each of

$$y_1(x) = x, \quad y_2(x) = \frac{1}{x}, \quad y_3(x) = x^2$$

into the differential equation, we find that, in each case, the equation is satisfied identically. Hence, each of the functions $y_i(x)$, $i = 1, 2, 3$ is a solution of the given differential equation. To show that these solutions are, in fact, linearly independent, we calculate the Wronskian (see Section 3.1). Expanding the determinant along the third row, we have

$$\begin{aligned} W(x) &= \begin{vmatrix} x & \frac{1}{x} & x^2 \\ 1 & -\frac{1}{x^2} & 2x \\ 0 & \frac{2}{x^3} & 2 \end{vmatrix} \\ &= -\frac{2}{x^3}\left(2x^2 - x^2\right) + 2\left(-\frac{x}{x^2} - \frac{1}{x}\right) \\ &= -\frac{2}{x^3}\left(x^2\right) + 2\left(-\frac{2}{x}\right) \\ &= -\frac{6}{x} \neq 0 \end{aligned}$$

Hence, since the Wronskian is non-zero, we deduce that the solutions $y_i(x)$, $i = 1, 2, 3$ are in fact linearly independent (see Note 3.1.1). It follows that (see Section 3.1) the general solution of the equation is given by

$$\begin{aligned} y(x) &= c_1 y_1(x) + c_2 y_2(x) + c_3 y_3(x) \\ &= c_1 x + \frac{c_2}{x} + c_3 x^2 \end{aligned}$$

MIDTERM EXAMINATION #5
SOLUTIONS

In what follows, c_i, $i = 1, 2, ...$ will denote arbitrary constants. Rules of logarithms and exponentials, a summary of the main techniques of integration as well as a *table of integrals* can be found in the Appendix.

1. Find the general solution.

(i) (5%) $\dfrac{1}{e^y}\dfrac{dy}{dt} = 1 + e^t - e^{-y} - e^{t-y}$

(ii) (15%) $\dfrac{dy}{dx} = \dfrac{6y^2}{x\left(2x^3 + y\right)}$

Solution.

Question 1.(i) Write the equation in the form

$$e^{-y}\frac{dy}{dt} = (1 + e^t)(1 - e^{-y})$$

to reveal that it is, in fact, a *separable* equation (see Section 2.1). Following the procedure in Section 2.1, we have

$$\begin{aligned} \underbrace{\int \frac{e^{-y}}{1 - e^{-y}} dy}_{\text{Let } u = e^{-y}} &= \int \left(1 + e^t\right) dt \\ \ln\left|1 - e^{-y}\right| &= t + e^t + c_1 \end{aligned}$$

which is, therefore, the general solution.

Question 1.(ii) This equation is particularly difficult to identify - it seems not to fit any of the categories discussed in Chapter 2. However, if we write the equation in the form

$$\frac{dx}{dy} = \frac{x(2x^3 + y)}{6y^2}$$

or

$$\frac{dx}{dy} - x\left(\frac{1}{6y}\right) = x^4 \frac{1}{3y^2} \qquad \text{(mt5.1)}$$

we can see that this is in fact a Bernoulli equation with x *as the dependent variable!* i.e. the roles of x and y in the theory for the Bernoulli equation of Section 2.6 have been reversed - otherwise, the theory and solution procedure is exactly the same. Following the procedure of Section 2.6 (with the roles of x and y reversed), we let

$$v(y) = x^{1-4} = x^{-3}$$

Then

$$\frac{dv}{dy} = -3x^{-4}\frac{dx}{dy}$$

so that the differential equation (mt5.1) becomes

$$\begin{aligned} -\frac{1}{3}\frac{dv}{dy} - \frac{1}{6y}v &= \frac{1}{3y^2} \\ \frac{dv}{dy} + \frac{1}{2y}v &= -\frac{1}{y^2} \end{aligned} \qquad \text{(mt5.2)}$$

which is linear in v. (see Section 2.5). An integrating factor is given by

$$\mu(y) = \exp\left(\int \frac{1}{2y}dy\right) = |y|^{\frac{1}{2}}$$

Choose

$$\mu(y) = y^{\frac{1}{2}}$$

The differential equation (mt5.2) may now be written in the form

$$\frac{d}{dy}\left(vy^{\frac{1}{2}}\right) = -y^{\frac{1}{2}} \cdot \frac{1}{y^2} = -y^{-\frac{3}{2}}$$

Integrating both sides with respect to y we obtain

$$\begin{aligned} vy^{\frac{1}{2}} &= 2y^{-\frac{1}{2}} + c_1 \\ v(y) &= \frac{2}{y} + c_1 y^{-\frac{1}{2}} \end{aligned}$$

Now let $v = x^{-3}$:

$$\begin{aligned} x^{-3} &= \frac{2}{y} + c_1 y^{-\frac{1}{2}} \\ \frac{1}{x^3} &= \frac{2 + c_1 y^{\frac{1}{2}}}{y} \end{aligned}$$

so that the general solution is given by

$$\frac{y - 2x^3}{x^3 y^{\frac{1}{2}}} = c_1$$

Question 2.(10%) Find all possible functions $g(x)$ such that the differential equation

$$\frac{dy}{dx} = -\frac{y \sin x}{g(x)}$$

is exact. Solve the differential equation for these functions $g(x)$.

Solution. The differential equation

$$\frac{dy}{dx} = -\frac{y \sin x}{g(x)}$$

or

$$\underbrace{y \sin x}_{M}\, dx + \underbrace{g(x)}_{N}\, dy = 0$$

is exact if and only if (see Section 2.3)

$$\frac{\partial M}{\partial y} = \frac{\partial N}{\partial x}$$

i.e.

$$\begin{aligned} \frac{\partial}{\partial y}(y \sin x) &= g'(x) \\ \sin x &= g'(x) \\ g(x) &= -\cos x + A \end{aligned}$$

Hence, the differential equation is exact for all functions $g(x)$ of the form $g(x) = -\cos x + A$, where A is an arbitrary constant. For such functions, the general solution of the differential equation is given by

$$F(x, y) = c_1$$

where the function $F(x, y)$ is given by

$$\frac{\partial F}{\partial x} = y \sin x \qquad \text{(mt5.3)}$$

$$\frac{\partial F}{\partial y} = g(x) = -\cos x + A \qquad \text{(mt5.4)}$$

From (mt5.3)

$$F(x, y) = -y \cos x + f(y)$$

where f is an arbitrary function of y. To satisfy (mt5.4), we require that

$$\begin{aligned} -\cos x + f'(y) &= -\cos x + A \\ f'(y) &= A \\ f(y) &= Ay + c_2 \end{aligned}$$

Hence,

$$F(x,y) = -y\cos x + Ay + c_2$$

and the general solution of the differential equation for functions g of the form specified above, is given by

$$\begin{aligned} F(x,y) &= c_1 \\ -y\cos x + Ay + c_2 &= c_1 \end{aligned}$$

or

$$-y\cos x + Ay = c_3$$

where $c_3 = c_1 - c_2$.

Question 3.(10%) Solve the initial value problem

$$(t-y)dt + (3t+y)dy = 0, \quad y(1) = 0$$

Solution. Write the differential equation in standard form.

$$\frac{dy}{dt} = \frac{y-t}{3t+y} = f(t,y)$$

The equation is most likely *homogeneous* (Section 2.2) - this is not difficult to confirm:

$$\begin{aligned} f(\lambda t, \lambda y) &= \frac{\lambda y - \lambda t}{3\lambda t + \lambda y} \\ &= \frac{y-t}{3t+y} \\ &= f(t,y) \end{aligned}$$

Hence, let $y = vt$ so that the differential equation becomes

$$\begin{aligned} v + t\frac{dv}{dt} &= \frac{vt-t}{3t+vt} \\ &= \frac{v-1}{3+v} \end{aligned}$$

Thus,

$$\begin{aligned} t\frac{dv}{dt} &= -\frac{2v+v^2+1}{3+v} \\ &= -\frac{(v+1)^2}{v+3} \end{aligned}$$

This equation is now *separable* (Section 2.1).

$$\begin{aligned} \int \frac{v+3}{(v+1)^2}dv &= -\int \frac{dt}{t} \\ \int [\frac{v+1}{(v+1)^2} + \frac{2}{(v+1)^2}]dv &= -\int \frac{dt}{t} \\ \int [\frac{1}{(v+1)} + \frac{2}{(v+1)^2}]dv &= -\ln|t| + c_1 \\ \ln|v+1| - 2(v+1)^{-1} &= -\ln|t| + c_1 \end{aligned}$$

Now let $v = \frac{y}{t}$, to obtain the general solution.

$$\begin{aligned} \ln\left|\frac{y}{t}+1\right| - \frac{2}{(\frac{y}{t}+1)} &= -\ln|t| + c_1 \\ \ln\left|t\left(\frac{y}{t}+1\right)\right| &= c_1 + \frac{2}{(\frac{y}{t}+1)} \\ \ln|y+t| &= c_1 + \frac{2t}{y+t} \end{aligned}$$

Apply the initial condition, $y(1) = 0$,

$$\begin{aligned} 0 &= c_1 + 2 \\ c_1 &= -2 \end{aligned}$$

Finally, the solution of the initial value problem is given by

$$\begin{aligned} \ln|y+t| &= 2\left[\frac{t}{y+t} - 1\right] \\ &= \frac{-2y}{y+t} \end{aligned}$$

Question 4.(15%) It is known that the differential equation

$$x^3y'' + xy' - y = 0 \qquad \text{(mt5.5)}$$

has a solution of the form $y = x^n$, for some positive integer n. Find such a solution and hence find the general solution of the inhomogeneous equation

$$x^3y'' + xy' - y = 1 \qquad \text{(mt5.6)}$$

Solution. Try $n = 1$ i.e. let $y_1(x) = x$. Substituting this into the homogeneous differential equation (mt5.5), we see that $y_1 = x$ is indeed a solution. This is sufficient information to enable us to attempt a solution of the inhomogeneous equation (mt5.6) by *reduction of order* (Section 2.4). Hence, we suggest that the general solution of (mt5.6) is given by $y(x) = vy_1 = vx$ where v is an unknown function of x to be determined.

$$\begin{aligned} y &= vx \\ y' &= v + xv' \\ y'' &= xv'' + 2v' \end{aligned}$$

Substitute into the differential equation (mt5.6).

$$\begin{aligned} x^3(xv'' + 2v') + x(v + xv') - vx &= 1 \\ x^4v'' + x^2(2x+1)v' &= 1 \end{aligned}$$

Let $w = v'$.

$$w' + \frac{2x+1}{x^2}w = \frac{1}{x^4}$$

This equation is linear in w and an integrating factor is given by

$$\begin{aligned} \mu(x) &= \exp\left(\int \frac{2x+1}{x^2}dx\right) \\ &= \exp\left(2\ln|x| - \frac{1}{x}\right) \\ &= x^2e^{-\frac{1}{x}} \end{aligned}$$

The differential equation in w now becomes

$$\frac{d}{dx}\left(wx^2e^{-\frac{1}{x}}\right) = \frac{e^{-\frac{1}{x}}}{x^2}$$

Integrating both sides with respect to x, we obtain

$$\begin{aligned} wx^2e^{-\frac{1}{x}} &= \underbrace{\int \frac{e^{-\frac{1}{x}}}{x^2}dx}_{\text{Let } u=-\frac{1}{x}} \\ wx^2e^{-\frac{1}{x}} &= e^{-\frac{1}{x}} + c_1 \\ w &= x^{-2} + \frac{c_1}{x^2}e^{\frac{1}{x}} \end{aligned}$$

Finally,

$$\begin{aligned}
v &= \int w dx \\
&= \int x^{-2} dx + c_1 \underbrace{\int \frac{e^{\frac{1}{x}}}{x^2} dx}_{\text{Let } u=\frac{1}{x}} \\
&= -\frac{1}{x} + c_1 \left(-e^{\frac{1}{x}}\right) + c_2
\end{aligned}$$

The general solution of the equation (mt5.6) is therefore given by

$$\begin{aligned}
y &= vx \\
&= x \left[-\frac{1}{x} + c_1 \left(-e^{\frac{1}{x}}\right) + c_2\right] \\
&= -1 + c_2 x - c_1 x e^{\frac{1}{x}}
\end{aligned}$$

5.(a)(5%) Is every separable equation exact ? Justify your answer with either a proof or a counter-example.

5.(b)(5%) If the Wronskian of two functions is identically zero, is it the case that these two functions must be linearly dependent ? Justify your answer with either a proof or a counter-example.

Solution

Question 5.(a) No. The fact that *not* every separable equation is exact is illustrated by the counter-example

$$\frac{dy}{dx} = \frac{y}{x}$$

or

$$xdy - ydx = 0$$

This differential equation is separable but not exact.

Question 5.(b) No. Consider the two functions

$$f_1(x) = x^2, \quad f_2(x) = \begin{cases} x^2, & x \geq 0, \\ 0, & x < 0 \end{cases}$$

Clearly, the Wronskian

$$W(f_1, f_2) = \begin{vmatrix} f_1 & f_2 \\ f_1' & f_2' \end{vmatrix}$$

is identically zero. However, if f_1 and f_2 are linearly dependent then there exist *constants* c_1 and c_2, *not both zero,* such that

$$c_1 f_1(x) + c_2 f_2(x) = 0 \text{ For all } x \tag{mt5.7}$$

However, when $x = 1$, (mt5.7) becomes

$$c_1 f_1(1) + c_2 f_2(1) = 0$$

which is equivalent to

$$c_1 + c_2 = 0 \tag{mt5.8}$$

Similarly, when $x = -1$,(mt5.7) becomes

$$c_1 f_1(-1) + c_2 f_2(-1) = 0$$

which is equivalent to

$$c_1 = 0$$

From (mt5.8) we obtain further that $c_2 = 0$, which contradicts the assumption that the functions f_1 and f_2 are linearly dependent. Thus, the functions f_1 and f_2 are linearly independent despite the fact that the Wronskian is identically zero.

Question 6.(10%) Solve the boundary value problem

$$\frac{d^2x}{dt^2} - 2\frac{dx}{dt} + 2x = e^t \cos 2t, \quad x'(0) = 0, \quad x(0) = 1$$

Solution. Since the differential equation is linear, the general solution $x(t)$ decomposes into two parts (see Section 3.1):

$$x(t) = x_c(t) + x_p(t)$$

For $x_c(t)$, the characteristic equation is given by

$$\begin{aligned} m^2 - 2m + 2 &= 0 \\ m &= 1 \pm i \end{aligned}$$

Hence, from Table 3.2.1,

$$x_c(t) = (c_1 \cos t + c_2 \sin t) e^t$$

For $x_p(t)$, using Table 3.3.1, try

$$x_p(t) = e^t (A\cos 2t + B\sin 2t)$$

Since $x_p(t)$ and $x_c(t)$ have no terms in common, there is no need to modify $x_p(t)$. To find the definite constants A and B, we substitute $x_p(t)$ into the differential equation.

$$\begin{aligned}
x_p(t) &= e^t (A\cos 2t + B\sin 2t) \\
x_p'(t) &= e^t (A\cos 2t + B\sin 2t) + e^t(-2A\sin 2t + 2B\cos 2t) \\
&= e^t[(A+2B)\cos 2t + (B-2A)\sin 2t] \\
x_p''(t) &= e^t[(A+2B)\cos 2t + (B-2A)\sin 2t - 2(A+2B)\sin 2t + 2(B-2A)\cos 2t] \\
&= e^t[(4B-3A)\cos 2t - (3B+4A)\sin 2t]
\end{aligned}$$

The differential equation becomes

$$\begin{aligned}
\frac{d^2x_p}{dt^2} - 2\frac{dx_p}{dt} + 2x_p &= e^t\cos 2t \\
-3e^t(A\cos 2t + B\sin 2t) &= e^t\cos 2t
\end{aligned}$$

Hence,

$$B = 0 \text{ and } A = -\frac{1}{3}$$

so that

$$x_p(t) = -\frac{1}{3}e^t\cos 2t$$

Finally, the general solution is given by

$$\begin{aligned}
x(t) &= x_c(t) + x_p(t) \\
&= (c_1\cos t + c_2\sin t)\,e^t - \frac{1}{3}e^t\cos 2t
\end{aligned}$$

The initial condition $x(0) = 1$ requires that

$$\begin{aligned}
1 &= c_1 - \frac{1}{3} \\
c_1 &= \frac{4}{3}
\end{aligned}$$

and $x'(0) = 0$ requires that

$$\begin{aligned}
0 &= c_1 + c_2 - \frac{1}{3} \\
0 &= \frac{4}{3} + c_2 - \frac{1}{3} \\
c_2 &= -1
\end{aligned}$$

The solution of the initial value problem is therefore

$$x(t) = \left(\frac{4}{3}\cos t - \sin t\right)e^t - 3e^t\cos 2t$$

Question 7.(10%) Investigate the solution of the boundary value problem

$$\frac{d^2y}{dx^2} + y = x^3, \quad y(0) = 0, \quad y(\pi) = 0$$

Solution. Since the differential equation is linear, the general solution $y(x)$ decomposes into two parts (see Section 3.1):

$$y(x) = y_c(x) + y_p(x)$$

For $y_c(x)$, the characteristic equation is given by

$$\begin{aligned} m^2 + 1 &= 0 \\ m &= \pm i \end{aligned}$$

Hence, from Table 3.2.1,

$$y_c(x) = c_1 \cos x + c_2 \sin x$$

For $y_p(x)$, using Table 3.3.1, try

$$y_p(x) = A + Bx + Cx^2 + Ex^3$$

Since $y_p(x)$ and $y_c(x)$ have no terms in common, there is no need to modify $y_p(x)$. To find the definite constants A, B, C and E, we substitute $y_p(x)$ into the differential equation.

$$\begin{aligned} y_p(x) &= A + Bx + Cx^2 + Ex^3 \\ y_p'(x) &= B + 2Cx + 3Ex^2 \\ y_p''(x) &= 2C + 6Ex \end{aligned}$$

The differential equation becomes

$$\begin{aligned} \frac{d^2y_p}{dx^2} + y_p &= x^3 \\ 2C + 6Ex + A + Bx + Cx^2 + Ex^3 &= x^3 \\ Ex^3 + Cx^2 + (B + 6E)x + A + 2C &= x^3 \end{aligned}$$

Hence,

$$\begin{aligned} A + 2C &= 0 \\ B + 6E &= 0 \\ C &= 0 \\ E &= 1 \end{aligned}$$

Solving this system leads to

$$A = 0, \quad B = -6, \quad C = 0 \text{ and } E = 1$$

so that

$$y_p(x) = -6x + x^3$$

Finally, the general solution is given by

$$\begin{aligned} y(x) &= y_c(x) + y_p(x) \\ &= c_1 \cos x + c_2 \sin x - 6x + x^3 \end{aligned}$$

The conditions $y(0) = 0, \; y(\pi) = 0$ lead to

$$\begin{aligned} 0 &= c_1 \\ 0 &= -6\pi + \pi^3 \; ? \end{aligned}$$

The last equation cannot possibly be satisfied. Hence the boundary value problem has no solution.

Question 8.(15%) Use a suitable substitution to solve

$$\left[\left(\frac{x}{y}\right)^2 \left(1 + e^{yx^{-1}}\right)^{-1} + \frac{y}{x}\right] dx - dy = 0, \quad y(1) = 0$$

Solution. If we write the equation in standard from, we can readily verify that the equation is, in fact, *homogeneous* (see Section 2.2). Noting the comment made in Note 2.2.2, we let $y = vx$. The differential equation now becomes

$$\begin{aligned} \frac{dy}{dx} &= \left(\frac{x}{y}\right)^2 \left(1 + e^{yx^{-1}}\right)^{-1} + \frac{y}{x} \\ v + x\frac{dv}{dx} &= \left(\frac{1}{v}\right)^2 (1 + e^v)^{-1} + v \\ x\frac{dv}{dx} &= \frac{1}{v^2(1 + e^v)} \\ v^2(1 + e^v)\, dv &= \frac{dx}{x} \end{aligned}$$

This equation is *separable* (see Section 2.1).

$$\begin{aligned} \int v^2 (1 + e^v)\, dv &= \int \frac{dx}{x} \\ \frac{v^3}{3} + \underbrace{\int v^2 e^v dv}_{Parts} &= \ln|x| + c_1 \\ \frac{v^3}{3} + e^v\left(v^2 - 2v + 2\right) &= \ln|x| + c_1 \end{aligned}$$

Let $v = \dfrac{y}{x}$.

$$\frac{y^3}{3x^3} + e^{\frac{y}{x}}[\left(\frac{y}{x}\right)^2 - 2\left(\frac{y}{x}\right) + 2] = \ln|x| + c_1$$

Apply the given boundary condition.

$$\begin{aligned} y(1) &= 0 \\ 2 &= c_1 \end{aligned}$$

Hence, the solution of the boundary value problem is given by

$$\frac{y^3}{3x^3} + e^{\frac{y}{x}}[\left(\frac{y}{x}\right)^2 - 2\left(\frac{y}{x}\right) + 2] = \ln|x| + 2$$

FINAL EXAMINATION #1

SOLUTIONS

In what follows, c_i, $i = 1, 2, ...$ will denote arbitrary constants. Rules of logarithms and exponentials, a summary of the main techniques of integration as well as a *table of integrals* can be found in the Appendix.

Question 1.(20%) Solve the following differential equation

$$x^2\frac{d^2y}{dx^2} + x\frac{dy}{dx} - y = x^2 \ln x, \quad x > 0$$

Solution. The equation is linear. This means that the general solution can be written in the form (see Section 3.1)

$$y(x) = y_c(x) + y_p(x)$$

For $y_c(x)$, we note that the corresponding homogeneous equation

$$x^2\frac{d^2y}{dx^2} + x\frac{dy}{dx} - y = 0 \tag{F1.1}$$

is of the Euler-Cauchy type. Following the procedure in Section 3.6, we make the substitution

$$x = e^t, \quad \frac{dy}{dx} = e^{-t}\frac{dy}{dt}, \quad \frac{d^2y}{dx^2} = e^{-2t}\left(\frac{d^2y}{dt^2} - \frac{dy}{dt}\right)$$

and the equation (F1.1) becomes

$$\begin{aligned} e^{2t}\left[e^{-2t}\left(\frac{d^2y}{dt^2} - \frac{dy}{dt}\right)\right] + e^t\left(e^{-t}\frac{dy}{dt}\right) - y &= 0 \\ \frac{d^2y}{dt^2} - y &= 0 \end{aligned}$$

This equation is now linear with *constant coefficients*. As in Section 3.2, its general solution is given by

$$y(t) = c_1e^t + c_2e^{-t}$$

or, in terms of the original variable x,

$$y(x) = c_1x + \frac{c_2}{x}$$

Hence,

$$y_c(x) = c_1 x + \frac{c_2}{x}$$

To find $y_p(x)$,we can use either Reduction of Order or Variation of Parameters. Since we have y_c, entirely, we use Variation of Parameters. From Section 3.5 we have that

$$y_p(x) = A(x)x + \frac{B(x)}{x}$$

where the unknown functions A and B are given by

$$\begin{bmatrix} x & \frac{1}{x} \\ 1 & -\frac{1}{x^2} \end{bmatrix} \begin{bmatrix} A' \\ B' \end{bmatrix} = \begin{bmatrix} 0 \\ \ln x \end{bmatrix}$$

Solving this matrix equation, we have

$$\begin{aligned} \begin{bmatrix} A' \\ B' \end{bmatrix} &= -\frac{x}{2} \begin{bmatrix} -\frac{1}{x^2} & -\frac{1}{x} \\ -1 & x \end{bmatrix} \begin{bmatrix} 0 \\ \ln x \end{bmatrix} \\ &= -\frac{x}{2} \begin{bmatrix} -\frac{\ln x}{x} \\ x \ln x \end{bmatrix} \end{aligned}$$

Hence,

$$\begin{aligned} A(x) &= \frac{1}{2} \int \ln x dx \\ B(x) &= -\frac{1}{2} \int x^2 \ln x dx \end{aligned}$$

Using integration by parts (or tables of integrals - see the Appendix) we have that

$$\begin{aligned} A(x) &= \frac{1}{2} (x \ln x - x) \\ B(x) &= -\frac{1}{18} (3x^3 \ln x - x^3) \end{aligned}$$

Note that, since we seek only a particular solution $y_p(x)$, we have set all constants of integration to zero. This means that

$$\begin{aligned} y_p(x) &= A(x)x + \frac{B(x)}{x} \\ &= \frac{1}{2}(x \ln x - x)\, x - \frac{(3x^3 \ln x - x^3)}{18x} \\ &= \frac{1}{2} x^2 \ln x - \frac{x^2}{2} - \frac{1}{18}(3x^2 \ln x - x^2) \\ &= \frac{1}{2} x^2 \ln x - \frac{1}{6} x^2 \ln x - \frac{4}{9} x^2 \\ &= \frac{1}{3} x^2 \ln x - \frac{4}{9} x^2 \end{aligned}$$

Finally, the general solution is given by

$$\begin{aligned} y(x) &= y_c(x) + y_p(x) \\ &= c_1 x + \frac{c_2}{x} + \frac{1}{3}x^2 \ln x - \frac{4}{9}x^2 \end{aligned}$$

Question 2.(10%) Use the method of *reduction of order* to find the general solution of the differential equation

$$\frac{d^2y}{dx^2} - 4\frac{dy}{dx} + 4y = \frac{e^{2x}}{x^2}, \quad x > 0$$

Solution. To use Reduction of Order we need one solution of the corresponding homogeneous equation. Since the equation is linear the general solution can be written in the form (see Section 3.1)

$$y(x) = y_c(x) + y_p(x)$$

For $y_c(x)$, we note that the corresponding homogeneous equation has characteristic equation (see Section 3.2)

$$\begin{aligned} m^2 - 4m + 4 &= 0 \\ (m-2)^2 &= 0 \\ m &= 2 \text{ (twice)} \end{aligned}$$

From Table 3.2.1, the general solution is given by

$$y(x) = (c_1 + c_2 x)\, e^{2x}$$

Hence,

$$y_c(x) = (c_1 + c_2 x)\, e^{2x}$$

To find $y_p(x)$ by reduction of order, we choose one of the solutions from y_c, for example, e^{2x} and let

$$\begin{aligned} y_p(x) &= ve^{2x} \\ y_p'(x) &= v'e^{2x} + 2ve^{2x} \\ y_p''(x) &= v''e^{2x} + 4v'e^{2x} + 4ve^{2x} \end{aligned}$$

where v is a function of x to be determined. Substituting into the inhomogeneous differential equation, we obtain

$$\begin{aligned} \frac{d^2y}{dx^2} - 4\frac{dy}{dx} + 4y &= \frac{e^{2x}}{x^2} \\ v''e^{2x} + 4v'e^{2x} + 4ve^{2x} - 4\left(v'e^{2x} + 2ve^{2x}\right) + 4ve^{2x} &= \frac{e^{2x}}{x^2} \\ v'' &= \frac{1}{x^2} \end{aligned}$$

Integrating with respect to x, we obtain (setting all constants of integration to zero)

$$\begin{aligned} v'(x) &= -\frac{1}{x} \\ v(x) &= -\ln x \end{aligned}$$

Hence,

$$y_p(x) = -e^{2x}\ln x$$

Finally, the general solution is given by

$$\begin{aligned} y(x) &= y_c(x) + y_p(x) \\ &= (c_1 + c_2 x)\, e^{2x} - e^{2x}\ln x \end{aligned}$$

Question 3.(10%) Determine the appropriate form for a particular solution of the differential equation

$$(D-1)^4\left(D^2+16\right)^2 y = xe^x + x\cos 4x$$

when using the method of undetermined coefficients. **Do not evaluate the coefficients.** Note that, here, $D^n \equiv \dfrac{d^n}{dx^n}$.

Solution. The form of the particular solution $y_p(x)$ is dependent on the complementary solution $y_c(x)$. In fact, the characteristic equation for the corresponding homogeneous equation is given by

$$\begin{aligned} (m-1)^4\left(m^2+16\right)^2 &= 0 \\ m &= 1 \text{ (4 times)},\ \pm 4i \text{ (twice)} \end{aligned}$$

Hence, from Table 3.2.1,

$$y_c(x) = \left(c_1 + c_2 x + c_3 x^2 + c_4 x^3\right) e^x + \left[(c_5 + c_6 x)\cos 4x + (c_7 + c_8 x)\sin 4x\right]$$

If we write

$$\begin{aligned} R(x) &= xe^x + x\cos 4x \\ &= R_1(x) + R_2(x) \end{aligned}$$

from Table 3.3.1, we suggest the following form for $y_p(x)$.

$$\begin{aligned} y_p(x) &= y_{p_1}(x) + y_{p_2}(x) \\ &= \underbrace{\underset{\substack{\uparrow \\ (*)}}{x^3}(Ax+B)\,e^x}_{y_{p_1}(x)} + \underbrace{\underset{\substack{\uparrow \\ (*)}}{x^2}\left[\,(C+Ex)\cos 4x + (F+Gx)\sin 4x\right]}_{y_{p_2}(x)} \end{aligned}$$

Here A, B, C, E, F and G are definite constants to be determined. Notice that each of the terms y_{p_1} and y_{p_2} is multiplied by the factor x^s where s is large enough to avoid any duplication between y_p and y_c, that is, $s = 3$ in y_{p_1} and $s = 2$ in y_{p_2}.

4. Find the Laplace transform of each of the following functions.

(i) (10%)

$$g(t) = \begin{cases} \cos t, & 0 < t < \pi, \\ 0, & \pi < t < 2\pi, \end{cases}$$

$g(t) = g(t + 2\pi), \ t \geq 0.$

(ii) (5%)

$$h(t) = e^t \cos 2t + t \sin t$$

Solution

Question 4.(i) The function $g(t)$ is periodic with period $p = 2\pi$. From Table 3.9.2

$$\begin{aligned} L[g(t)] &= \frac{1}{1 - e^{-2\pi s}} \int_0^{2\pi} e^{-st} g(t)\, dt \\ &= \frac{1}{1 - e^{-2\pi s}} \int_0^{\pi} e^{-st} \cos t dt \end{aligned}$$

Using a table of integrals (Appendix) or integration by parts, we have that

$$\begin{aligned} \int_0^{\pi} e^{-st} \cos t dt &= \left[\frac{e^{-st}}{s^2 + 1} (-s \cos t + \sin t) \right]_{t=0}^{t=\pi} \\ &= \frac{1}{s^2 + 1} \left[e^{-\pi s} (-s(-1)) - (-s) \right] \\ &= \frac{s(1 + e^{-\pi s})}{s^2 + 1} \end{aligned}$$

Hence,

$$\begin{aligned} L[g(t)] &= \frac{1}{1 - e^{-2\pi s}} \int_0^{\pi} e^{-st} \cos t dt \\ &= \frac{s(1 + e^{-\pi s})}{(1 - e^{-2\pi s})(s^2 + 1)} \end{aligned}$$

Question 4.(ii)

$$\begin{aligned} L[h(t)] &= L\left[e^t \cos 2t + t \sin t\right] \\ &= L\left[e^t \cos 2t\right] + L[t \sin t] \end{aligned}$$

From Table 3.9.2,

$$L[e^{at}y\,(t)] = f\,(s-a)$$

where $f\,(s) = L\,[y(t)]$. Hence,

$$L\left[e^{t}\cos 2t\right] = f_1\,(s-1)$$

where,

$$f_1\,(s) = L\,[\cos 2t] = \frac{s}{s^2+4}$$

so that

$$f_1\,(s-1) = \frac{s-1}{(s-1)^2+4}$$

Thus,

$$\begin{aligned} L\left[e^{t}\cos 2t\right] &= f_1\,(s-1) \\ &= \frac{s-1}{(s-1)^2+4} \\ &= \frac{s-1}{s^2-2s+5} \end{aligned}$$

From Table 3.9.1, we have,

$$L\,[t\sin t] = \frac{2s}{(s^2+1)^2}\,.$$

Finally,

$$\begin{aligned} L\,[h\,(t)] &= L\left[e^{t}\cos 2t\right] + L\,[t\sin t] \\ &= \frac{s-1}{s^2-2s+5} + \frac{2s}{(s^2+1)^2} \end{aligned}$$

Question 5.(15%) Solve the following initial value problem using the method of Laplace transforms.

$$\begin{aligned} \frac{d^2x}{dt^2} + x &= \begin{cases} 2, & 0 \le t < \frac{\pi}{2}, \\ 0, & t \ge \frac{\pi}{2}, \end{cases} \\ x\,(0) &= 0, \quad \frac{dx}{dt}\,(0) = 2 \end{aligned}$$

Solution. We write the differential equation in the form

$$\frac{d^2x}{dt^2} + x = 2\left[1 - \alpha\left(t - \frac{\pi}{2}\right)\right]$$

where α is the *Heaviside function* defined in Table 3.9.2. Applying the Laplace transform to both sides of the differential equation (and using Table 3.9.2), we obtain

$$\begin{aligned} L\left[\frac{d^2x}{dt^2}\right] + L[x] &= 2L\left[1 - \alpha\left(t - \frac{\pi}{2}\right)\right] \\ L\left[\frac{d^2x}{dt^2}\right] + L[x] &= 2L[1] - 2L[\alpha\left(t - \frac{\pi}{2}\right)] \\ s^2 L[x(t)] - sx(0) - \frac{dx}{dt}(0) + L[x(t)] &= \frac{2}{s}\left[1 - e^{-\frac{\pi}{2}s}\right] \end{aligned}$$

Next, apply the given initial conditions, $x(0) = 0$, $\frac{dx}{dt}(0) = 2$.

$$\begin{aligned} s^2 L[x(t)] - 2 + L[x(t)] &= \frac{2}{s}\left[1 - e^{-\frac{\pi}{2}s}\right] \\ L[x(t)](s^2+1) &= 2 + \frac{2}{s}\left[1 - e^{-\frac{\pi}{2}s}\right] \end{aligned}$$

Solve for $x(t)$.

$$x(t) = 2\{L^{-1}\left[\frac{1}{s^2+1}\right] + L^{-1}\left[\frac{1}{s(s^2+1)}\right] - L^{-1}\left[\frac{e^{-\frac{\pi}{2}s}}{s(s^2+1)}\right]\} \qquad \text{(F1.2)}$$

Now, using partial fractions and Table 3.9.1,

$$\begin{aligned} L^{-1}\left[\frac{1}{s(s^2+1)}\right] &= L^{-1}\left[\frac{1}{s} - \frac{s}{(s^2+1)}\right] \\ &= 1 - \cos t \end{aligned}$$

Using Table 3.9.2,

$$L^{-1}\left[\frac{e^{-\frac{\pi}{2}s}}{s(s^2+1)}\right] = \alpha\left(t - \frac{\pi}{2}\right) x_1\left(t - \frac{\pi}{2}\right)$$

where

$$\begin{aligned} x_1(t) &= L^{-1}\left[\frac{1}{s(s^2+1)}\right] \\ &= 1 - \cos t \ \text{ (as above)} \end{aligned}$$

Hence,

$$\begin{aligned} x_1\left(t - \frac{\pi}{2}\right) &= 1 - \cos\left(t - \frac{\pi}{2}\right) \\ &= 1 - \sin t \end{aligned}$$

and

$$\begin{aligned} L^{-1}\left[\frac{e^{-\frac{\pi}{2}s}}{s\left(s^2+1\right)}\right] &= \alpha\left(t-\frac{\pi}{2}\right)x_1\left(t-\frac{\pi}{2}\right) \\ &= \alpha\left(t-\frac{\pi}{2}\right)(1-\sin t) \end{aligned}$$

Finally, from (F1.2),

$$\begin{aligned} x\left(t\right) &= 2\{L^{-1}\left[\frac{1}{s^2+1}\right]+L^{-1}\left[\frac{1}{s\left(s^2+1\right)}\right]-L^{-1}\left[\frac{e^{-\frac{\pi}{2}s}}{s\left(s^2+1\right)}\right]\} \\ &= 2\{\sin t+1-\cos t-\alpha\left(t-\frac{\pi}{2}\right)(1-\sin t)\} \end{aligned}$$

Question 6.(10%) Use the convolution theorem and the method of Laplace transforms to solve

$$\frac{d^2y}{dx^2}+6\frac{dy}{dx}+9y=H\left(x\right), \quad y\left(0\right)=0, \quad y'\left(0\right)=1$$

where $H\left(x\right)$ is a known function of x.

Solution. Applying the Laplace transform to the differential equation yields

$$\begin{aligned} L[\frac{d^2y}{dx^2}+6\frac{dy}{dx}+9y] &= L[H\left(x\right)] \\ L[\frac{d^2y}{dx^2}]+6L[\frac{dy}{dx}]+9L[y] &= L[H\left(x\right)] \\ s^2L\left[y\left(x\right)\right]-sy\left(0\right)-y'\left(0\right)+6[sL[y\left(x\right)]-y\left(0\right)]+9L[y\left(x\right)] &= L[H\left(x\right)] \\ s^2L[y\left(x\right)]-1+6sL[y\left(x\right)]+9L[y\left(x\right)] &= L[H\left(x\right)] \end{aligned}$$

Solve for $L[y\left(x\right)]$, to obtain

$$\begin{aligned} L[y\left(x\right)] &= \frac{1}{s^2+6s+9}\left(1+L[H\left(x\right)]\right) \\ &= \frac{1}{\left(s+3\right)^2}+\frac{L[H\left(x\right)]}{\left(s+3\right)^2} \\ &= \frac{1}{\left(s+3\right)^2}+\frac{f_2\left(s\right)}{\left(s+3\right)^2} \end{aligned}$$

where, $f_2\left(s\right)=L[H(x)]\left(s\right)$. Next, take the inverse Laplace transform to obtain $y\left(x\right)$.

$$y\left(x\right)=L^{-1}[\frac{1}{\left(s+3\right)^2}]+L^{-1}\left[\frac{f_2\left(s\right)}{\left(s+3\right)^2}\right]$$

From Tables 3.9.1 and 3.9.2 (Convolution Theorem),

$$y(x) = xe^{-3x} + \int_0^x y_1(\beta)\, y_2(x-\beta)\, d\beta$$

where

$$\begin{aligned} y_1(\beta) &= L^{-1}[\frac{1}{(s+3)^2}](\beta) \\ &= \beta e^{-3\beta} \end{aligned}$$

and

$$\begin{aligned} y_2(x) &= L^{-1}[f_2(s)](x) \\ &= H(x) \end{aligned}$$

so that

$$y_2(x-\beta) = H(x-\beta)$$

Hence,

$$\begin{aligned} y(x) &= xe^{-3x} + \int_0^x y_1(\beta)\, y_2(x-\beta)\, d\beta \\ &= xe^{-3x} + \int_0^x \beta e^{-3\beta} H(x-\beta)\, d\beta \end{aligned}$$

Question 7.(20%) Find the general solution in a series about $x = 0$. Give a region of validity for your solution (you must justify your conclusion).

$$\left(4 - x^2\right) y'' - 2xy' + 2y = 0 \qquad \text{(F1.3)}$$

Solution. From Section 3.8, it is clear that $x = 0$ is an ordinary point and that there are singular points at $x = \pm 2$. Hence, in accordance with the theory of series solutions about an ordinary point (Theorem 3.8.3), the equation (F1.3) has two linearly independent series solutions of the form

$$y(x) = \sum_{n=0}^{\infty} b_n x^n$$

(b_n are constants to be determined) valid (at least) for $|x| < 2$. Noting that

$$\begin{aligned} y(x) &= \sum_{n=0}^{\infty} b_n x^n \\ y'(x) &= \sum_{n=1}^{\infty} n b_n x^{n-1} \\ y''(x) &= \sum_{n=2}^{\infty} n(n-1) b_n x^{n-2} \end{aligned}$$

(F1.3) becomes

$$\sum_{n=2}^{\infty} n(n-1) b_n (4x^{n-2} - x^n) - 2\sum_{n=1}^{\infty} nb_n x^n + 2\sum_{n=0}^{\infty} b_n x^n = 0$$

Shifting the summation in the first two sums on the left-hand side of this equation $(n \to n+2)$, we obtain

$$\begin{aligned}
4\sum_{n=0}^{\infty} (n+2)(n+1) b_{n+2} x^n - \sum_{n=0}^{\infty} n(n-1) b_n x^n - 2\sum_{n=0}^{\infty} nb_n x^n + 2\sum_{n=0}^{\infty} b_n x^n &= 0 \\
\sum_{n=0}^{\infty} (4(n+2)(n+1) b_{n+2} - n(n-1) b_n - 2nb_n + 2b_n) x^n &= 0 \\
\sum_{n=0}^{\infty} (4(n+2)(n+1) b_{n+2} - (n-1)(n+2) b_n) x^n &= 0
\end{aligned}$$

It follows that a recurrence relation for the coefficients b_n is given by

$$\begin{aligned}
b_{n+2} &= \frac{(n-1)(n+2)}{4(n+2)(n+1)} b_n, \quad n \geq 0 \\
&= \frac{n-1}{4(n+1)} b_n, \quad n \geq 0 \qquad \text{(F1.4)}
\end{aligned}$$

Let b_0 and b_1 be arbitrary constants. From (F1.4),

$$\begin{aligned}
b_0 &= b_0 & b_1 &= b_1 \\
b_2 &= \tfrac{-1b_0}{4\cdot 1} & b_3 &= 0 \\
b_4 &= \tfrac{1\cdot b_2}{4\cdot 3} = \tfrac{-1\cdot 1b_0}{4^2\cdot 3} & b_5 &= 0 \\
b_6 &= \tfrac{3\cdot b_4}{4\cdot 5} = \tfrac{-1\cdot 1\cdot 3b_0}{4^3\cdot 1\cdot 3\cdot 5} & b_7 &= 0 \\
&\vdots & &\vdots \\
&\vdots & &\vdots \\
b_{2n} &= \tfrac{-1\cdot 1\cdot 3\cdot 5\cdot 7 \ldots (2n-3)}{4^n\cdot 1\cdot 3\cdot \ldots (2n-3)(2n-1)} b_0\,, \quad n \geq 0 & &\vdots \\
b_{2n} &= -\tfrac{1}{4^n(2n-1)} b_0, \quad n \geq 0 & b_{2n+1} &= 0, \quad n \geq 1
\end{aligned}$$

Hence, the general solution of (F1.3) is given by

$$\begin{aligned}
y(x) &= \sum_{n=0}^{\infty} b_n x^n \\
&= b_0 + b_1 x + \sum_{n=1}^{\infty} b_{2n} x^{2n} + \sum_{n=1}^{\infty} b_{2n+1} x^{2n+1} \\
&= b_0 + b_1 x - \sum_{n=1}^{\infty} \frac{1}{4^n (2n-1)} b_0 x^{2n}
\end{aligned}$$

$$= b_0\left(1-\sum_{n=1}^{\infty}\frac{1}{4^n\,(2n-1)}x^{2n}\right)+b_1x$$

$$= -b_0\sum_{n=0}^{\infty}\frac{1}{4^n\,(2n-1)}x^{2n}+b_1x$$

This solution is valid, at least, for $|x| < 2$ (certainly 'near $x = 0$' as required). This minimum region of validity follows from the fact that the nearest singular points are at ± 2 (see Theorem 3.8.3). The region of validity may be even larger (the resulting series solution may have a larger radius of convergence).

FINAL EXAMINATION #2
SOLUTIONS

In what follows, c_i, $i = 1, 2, ...$ will denote arbitrary constants. Rules of logarithms and exponentials, a summary of the main techniques of integration as well as a *table of integrals* can be found in the Appendix.

Question 1. (10%) Solve the following nonlinear ordinary differential equation.

$$y\frac{d^2y}{dx^2} - \left(\frac{dy}{dx}\right)^2 = 0, \quad y > 0$$

Solution. This is one of the class of *Special Types of ODEs* considered in Section 3.7. Since the independent variable x does not appear explicitly, we let $y' = p$ and $y'' = p\frac{dp}{dy}$. The differential equation becomes

$$\begin{aligned} yp\frac{dp}{dy} - p^2 &= 0 \\ p\left(y\frac{dp}{dy} - p\right) &= 0 \\ p &= 0 \quad \text{or} \quad \frac{dp}{dy} = \frac{p}{y} \end{aligned}$$

If $p = 0$,

$$\begin{aligned} p &= \frac{dy}{dx} = 0 \\ y &= c_1 \end{aligned}$$

If $p \neq 0$, we have the equation

$$\frac{dp}{dy} = \frac{p}{y}$$

which is *separable* (see Section 2.1).

$$\begin{aligned} \int \frac{dp}{p} &= \int \frac{dy}{y} \\ \ln|p| &= \ln|y| + c_1 \\ \ln|p| &= \ln Ay \end{aligned}$$

since $y > 0$ and where $c_1 = \ln A$, $A > 0$. Thus,

$$\begin{aligned} |p| &= Ay \\ p &= By, \quad B = \pm A \end{aligned}$$

Next, let $p = \dfrac{dy}{dx}$ and obtain the following separable equation.

$$\begin{aligned}
\frac{dy}{dx} &= By \\
\int \frac{dy}{y} &= \int B dx \\
\ln|y| &= Bx + c_2 \\
|y| &= c_3 e^{Bx}, \quad c_3 = e^{c_2} \\
y &= c_4 e^{Bx}, \quad c_4 = \pm c_3
\end{aligned}$$

Finally, the general solution is given by

$$y(x) = c_4 e^{Bx}$$

Question 2. (15%) For the following homogeneous differential equation find a solution of the form x^n, where n is a positive integer.

$$(x^2 - 1)\frac{d^2y}{dx^2} - 2x\frac{dy}{dx} + 2y = 0$$

Use this solution to find the general solution of the following corresponding inhomogeneous equation

$$(x^2 - 1)\frac{d^2y}{dx^2} - 2x\frac{dy}{dx} + 2y = \left(x^2 - 1\right)^2$$

Solution. For a solution (of the *homogeneous* equation) of the form $y = x^n$, we first try $y = x$.

$$\begin{aligned}
(x^2 - 1)\frac{d^2y}{dx^2} - 2x\frac{dy}{dx} + 2y &= 0 \\
\left(x^2 - 1\right)(0) - 2x(1) + 2x &= 0 \\
0 &= 0
\end{aligned}$$

Hence, $y = x$ is indeed a solution of the *homogeneous* equation. To solve the *inhomogeneous* equation, we use the known solution $y = x$ of the homogeneous equation in a *reduction of order* process (see Section 3.4). Let v be an unknown function of x and suppose that the general solution of the *inhomogeneous* equation

$$(x^2 - 1)\frac{d^2y}{dx^2} - 2x\frac{dy}{dx} + 2y = \left(x^2 - 1\right)^2$$

is given by $y = vx$. Then

$$\begin{aligned}
y &= vx \\
y' &= xv' + v \\
y'' &= xv'' + 2v'
\end{aligned}$$

Substituting into the differential equation yields

$$\begin{aligned}\left(x^2-1\right)\left(xv''+2v'\right)-2x\left(xv'+v\right)+2xv &= \left(x^2-1\right)^2\\ \left(x^2-1\right)xv''+\ \left(2x^2-2-2x^2\right)v' &= \left(x^2-1\right)^2\\ v''-\frac{2}{x\left(x^2-1\right)}v' &= \frac{x^2-1}{x}\end{aligned}$$

Let $w=v'$ to obtain the following first order equation, *linear* in w.

$$w'-\frac{2}{x\left(x^2-1\right)}w=\frac{x^2-1}{x}$$

Following the procedure in Section 2.5, an integrating factor for this equation is given by

$$\begin{aligned}\mu(x) &= \exp\left(-\int\frac{2}{x\left(x^2-1\right)}dx\right)\\ &= \exp\left(\int\underbrace{\left[\frac{2}{x}-\frac{1}{x-1}-\frac{1}{1+x}\right]}_{\text{Partial Fractions - Appendix}}dx\right)\\ &= \exp\left(\ln x^2-\ln|x-1|-\ln|x+1|\right)\\ &= \exp\left(\ln\frac{x^2}{|x-1|\,|x+1|}\right)\\ &= \frac{x^2}{|x-1|\,|x+1|}\end{aligned}$$

We choose

$$\begin{aligned}\mu\left(x\right) &= \frac{x^2}{\left(x-1\right)\left(x+1\right)}\\ &= \frac{x^2}{x^2-1}\end{aligned}$$

and write the differential equation in w as

$$\begin{aligned}\frac{d}{dx}\left(w\cdot\frac{x^2}{x^2-1}\right) &= \frac{x^2}{x^2-1}\cdot\frac{x^2-1}{x}\\ &= x\end{aligned}$$

Integrate both sides with respect to x :

$$w\cdot\frac{x^2}{x^2-1}=\frac{x^2}{2}+c_1$$

Thus,

$$\begin{aligned} w(x) &= v'(x) \\ &= \frac{x^2-1}{2} + c_1 \frac{x^2-1}{x^2} \end{aligned}$$

Integrate again with respect to x to get $v(x)$.

$$\begin{aligned} v(x) &= \int \left(\frac{x^2-1}{2} + c_1 \frac{x^2-1}{x^2} \right) dx \\ &= \int \left(\frac{x^2}{2} - \frac{1}{2} + c_1 - \frac{c_1}{x^2} \right) dx \\ &= \frac{x^3}{6} - \frac{x}{2} + c_1 (x + \frac{1}{x}) + c_2 \end{aligned}$$

Finally, the general solution of the inhomogeneous equation is given by

$$\begin{aligned} y &= xv \\ &= x \left(\frac{x^3}{6} - \frac{x}{2} + c_1 (x + \frac{1}{x}) + c_2 \right) \\ &= \frac{x^4}{6} - \frac{x^2}{2} + c_1 \left(x^2 + 1 \right) + c_2 x \end{aligned}$$

3.

(i) (10%) Find

$$L^{-1} \left[\frac{5s+3}{s^2+4s+5} \right]$$

(ii) (10%) Determine

$$L^{-1} \left[\frac{s}{(s+3)^5 (s^2+16)} \right]$$

in the form of an integral (Do not evaluate the integral).

Solution

Question 3.(i) To find

$$L^{-1} \left[\frac{5s+3}{s^2+4s+5} \right]$$

complete the square in the denominator and write the expression as a function entirely of the resulting square:

$$\begin{aligned} L^{-1} \left[\frac{5s+3}{s^2+4s+5} \right] &= L^{-1} \left[\frac{5s+3}{(s+2)^2+1} \right] \\ &= L^{-1} \left[\frac{5(s+2)-7}{(s+2)^2+1} \right] \end{aligned}$$

Let

$$f(s) = \frac{5s-7}{s^2+1}$$

Then it is clear that we are required to find

$$L^{-1}[f(s+2)]$$

From Table 3.9.2,

$$\begin{aligned} L^{-1}[f(s+2)] &= e^{-2t}y(t) \\ & \quad e^{-2t}L^{-1}[f(s)] \\ &= e^{-2t}L^{-1}\left[\frac{5s-7}{s^2+1}\right] \\ &= e^{-2t}\left(5L^{-1}\left[\frac{s}{s^2+1}\right] - 7L^{-1}\left[\frac{1}{s^2+1}\right]\right) \end{aligned}$$

Now use Table 3.9.1.

$$\begin{aligned} L^{-1}\left[\frac{5s+3}{s^2+4s+5}\right] &= L^{-1}[f(s+2)] \\ &= e^{-2t}\left(5L^{-1}\left[\frac{s}{s^2+1}\right] - 7L^{-1}\left[\frac{1}{s^2+1}\right]\right) \\ &= e^{-2t}(5\cos t - 7\sin t) \end{aligned}$$

Question 3.(ii) Write

$$L^{-1}\left[\frac{s}{(s+3)^5(s^2+16)}\right] = L^{-1}[f_1(s) f_2(s)]$$

where

$$\begin{aligned} f_1(s) &= \frac{1}{(s+3)^5} \\ f_2(s) &= \frac{s}{s^2+16} \end{aligned}$$

The Convolution Theorem (see Table 3.9.2) now states that

$$L^{-1}[f_1(s) f_2(s)] = \int_0^t y_1(\beta)\, y_2(t-\beta)\, d\beta$$

where

$$\begin{aligned} y_1(t) &= L^{-1}[f_1(s)] \\ &= L^{-1}\left[\frac{1}{(s+3)^5}\right] \\ &= e^{-3t}L^{-1}\left[\frac{1}{s^5}\right] && \text{(using Table 3.9.1)} \\ &= e^{-3t}\frac{t^4}{4!} && \text{(using Table 3.9.1)} \end{aligned}$$

Similarly,

$$\begin{aligned} y_2(t) &= L^{-1}[f_2(s)] \\ &= L^{-1}\left[\frac{s}{s^2+16}\right] \\ &= \cos 4t \qquad \text{(using Table 3.9.1)} \end{aligned}$$

Finally,

$$\begin{aligned} L^{-1}\left[\frac{s}{(s+3)^5(s^2+16)}\right] &= L^{-1}[f_1(s)\,f_2(s)] \\ &= \int_0^t y_1(\beta)\,y_2(t-\beta)\,d\beta \\ &= \int_0^t e^{-3\beta}\frac{\beta^4}{4!}\cos 4(t-\beta)d\beta \end{aligned}$$

4.

(a) (10%) Find the general solution of the equation

$$y'' - 6y' + 13y = 15\cos 2x \qquad \text{(F2.1)}$$

(b) (5%) If the left-hand side of (F2.1) is replaced by $y'' + 4y$, what would be the corresponding form of particular solution y_p if the method of undetermined coefficients is used to find the general solution?

Solution

Question 4.(a) Since the equation (F2.1) is linear, its general solution $y(x)$ can be split into two parts (see Section 3.1):

$$y(x) = y_c(x) + y_p(x)$$

For $y_c(x)$, the characteristic equation is

$$\begin{aligned} m^2 - 6m + 13 &= 0 \\ m &= 3 \pm 2i \end{aligned}$$

From Table 3.2.1,

$$y_c(x) = e^{3x}(c_1\cos 2x + c_2\sin 2x)$$

For $y_p(x)$, from Table 3.3.1, we suggest

$$y_p(x) = A\cos 2x + B\sin 2x$$

where A and B are definite constants to be determined. Note that there is no need to multiply our initial guess for $y_p(x)$ by a factor of x^s since it contains no terms in common with y_c. To find A and B, we substitute $y_p(x)$ into the differential equation (F2.1).

$$\begin{aligned} y_p'' - 6y_p' + 13y_p &= 15\cos \\ (-4A\cos 2x - 4B\sin 2x) - 6(-2A\sin 2x + 2B\cos 2x) + 13(A\cos 2x + B\sin 2x) &= 15\cos \\ (9A - 12B)\cos 2x + (9B + 12A)\sin 2x &= 15\cos \end{aligned}$$

We obtain the following system of linear equations for A and B.

$$\begin{aligned} 9A - 12B &= 15 \\ 12A + 9B &= 0 \\ A &= \frac{3}{5}, B = -\frac{4}{5} \end{aligned}$$

Hence,

$$y_p(x) = \frac{1}{5}(3\cos 2x - 4\sin 2x)$$

Finally, the general solution of (F2.1) is given by

$$\begin{aligned} y(x) &= y_c(x) + y_p(x) \\ &= e^{3x}(c_1\cos 2x + c_2\sin 2x) + \frac{1}{5}(3\cos 2x - 4\sin 2x) \end{aligned}$$

Question 4.(b) If the left-hand side of (F2.1) is replaced by $y'' + 4y$, the corresponding $y_c(x)$ would be

$$y_c(x) = c_1\cos 2x + c_2\sin 2x$$

Hence, from Table 3.3.1, the corresponding form of $y_p(x)$ would be

$$y_p(x) = \underset{\substack{\uparrow \\ (*)}}{x}(A\cos 2x + B\sin 2x)$$

The additional factor 'x' is necessary to avoid terms in common with the (new) $y_c(x)$.

Question 5. (15%) Solve the following initial value problem using the method of Laplace transforms.

$$\frac{d^2y}{dx^2} + 4\frac{dy}{dx} + 4y = e^{-2x}, \quad y(0) = 0, \quad y'(0) = 1$$

Solution. First take Laplace transforms of both sides of the differential equation.

$$L[y''] + 4L[y'] + 4L[y] = L\left[e^{-2x}\right]$$

Denoting $L[y(x)]$ by $f(s)$ and using Table 3.9.2, we obtain

$$s^2 f(s) - sy(0) - y'(0) + 4sf(s) - 4y(0) + 4f(s) = L\left[e^{-2x}\right]$$

Apply the given initial conditions and use Table 3.9.1.

$$\begin{aligned} s^2 f(s) - 1 + 4sf(s) + 4f(s) &= \frac{1}{s+2} \\ f(s)\left[s^2 + 4s + 4\right] &= 1 + \frac{1}{s+2} \\ f(s) &= \frac{1}{(s+2)^2} + \frac{1}{(s+2)^3} \end{aligned}$$

Inverting the transform, from Table 3.9.2, we obtain

$$\begin{aligned} y(x) &= L^{-1}\left[\frac{1}{(s+2)^2}\right] + L^{-1}\left[\frac{1}{(s+2)^3}\right] \\ &= e^{-2x} L^{-1}\left[\frac{1}{s^2}\right] + e^{-2x} L^{-1}\left[\frac{1}{s^3}\right] \end{aligned}$$

From Table 3.9.1, we have

$$y(x) = e^{-2x}\left(x + \frac{x^2}{2}\right)$$

Question 6. (25%) Find the general solution of the following differential equation in terms of series centered at $x = 0$.

$$2x^2(1-x)\frac{d^2y}{dx^2} - x(1+x)\frac{dy}{dx} + (1+x)y = 0, \qquad x > 0 \qquad \text{(F2.2)}$$

Solution. From Section 3.8, $x = 0$ is an *ordinary* point of the differential equation if the expressions $P(x)$ and $Q(x)$ in (3.8.4) of Chapter 3 are *both* analytic at $x = 0$. In fact, here

$$\begin{aligned} P(x) &= \frac{-x(1+x)}{2x^2(1-x)} \\ Q(x) &= \frac{1+x}{2x^2(1-x)} \end{aligned}$$

It is clear that neither $P(x)$ nor $Q(x)$ are analytic at $x = 0$ so that $x = 0$ is a *singular* point of the differential equation. According to the theory in Section 3.8, $x = 0$ is a *regular singular point* if $p(x) = xP(x)$ and $q(x) = x^2Q(x)$ are *both* analytic at $x = 0$. In fact, here,

$$\begin{aligned} p(x) &= xP(x) = -\frac{1+x}{2(1-x)} \\ q(x) &= x^2Q(x) = \frac{1+x}{2(1-x)} \end{aligned}$$

which are indeed *both* analytic at $x = 0$. Hence, $x = 0$ is a *regular singular point* and, by Theorem 3.8.6, for $x > 0$, the differential equation has at least one Frobenius series solution (corresponding to the larger root of the indicial equation (3.8.13)). To see which form a second linearly independent solution will take, we examine the roots of the indicial equation (3.8.13) .

$$\begin{aligned} r(r-1) + p(0)r + q(0) &= 0 \\ r\,(r-1) - \frac{1}{2}r + \frac{1}{2} &= 0 \\ (2r-1)\,(r-1) &= 0 \\ r_1 &= 1, \quad r_2 = \frac{1}{2} \end{aligned}$$

The roots of the indicial equation differ by a *non-integer.* Hence, by Theorem 3.8.6, there exist two linearly independent Frobenius series solutions of the form

$$\begin{aligned} y_1(x) &= \sum_{n=0}^{\infty} a_n x^{n+r_1} \quad , \qquad a_0 \neq 0 \\ y_2(x) &= \sum_{n=0}^{\infty} b_n x^{n+r_2} \quad , \qquad b_0 \neq 0 \end{aligned}$$

These solutions will be valid, *at least,* in $0 < x < 1$ (since the nearest singular point of the functions $p\,(x)$ and $q\,(x)$ is at $x = 1$ - see Theorem 3.8.6). To find the coefficients a_n and b_n, we assume a (generic) series solution of the form

$$y\,(x) = \sum_{n=0}^{\infty} c_n x^{n+r} \quad , \qquad c_0 \neq 0, \ \ r = \frac{1}{2}, 1$$

Substituting into (F2.2), we obtain

$$\begin{aligned} & 2\sum_{n=0}^{\infty} c_n\,(n+r)\,(n+r-1)\,x^{n+r} - 2\sum_{n=0}^{\infty} c_n\,(n+r)\,(n+r-1)\,x^{n+r+1} \\ & -\sum_{n=0}^{\infty} c_n\,(n+r)\,x^{n+r+1} - \sum_{n=0}^{\infty} c_n\,(n+r)\,x^{n+r} \\ & +\sum_{n=0}^{\infty} c_n x^{n+r+1} + \sum_{n=0}^{\infty} c_n x^{n+r} \\ = \ & 0 \end{aligned}$$

Rearranging, we find that

$$\sum_{n=0}^{\infty} [(n+r)\,(2n+2r-3)+1] c_n x^{n+r} - \sum_{n=0}^{\infty} [(n+r)\,(2n+2r-1)-1] c_n x^{n+r+1} = 0$$

Simplifying and shifting summation index in the second sum ($n \to n-1$), we obtain

$$\sum_{n=0}^{\infty}[2(n+r)-1][n+r-1]c_n x^{n+r} - \sum_{n=1}^{\infty}[2(n+r)-1][n+r-2]c_{n-1}x^{n+r} = 0$$

Equate coefficients of x^n for different values of n :

$$n = 0: \quad (2r-1)(r-1)c_0 = 0, \quad c_0 \neq 0 \quad \text{(indicial equation)}$$

$$n \geq 1: \quad c_n(r) = \frac{[2(n+r)-1](n+r-2)}{[2(n+r)-1][n+r-1]}c_{n-1}$$

The final equation gives the recurrence relation for the coefficients c_n :

$$c_n(r) = \frac{n+r-2}{n+r-1}c_{n-1}, \quad n \geq 1$$

Let $r = r_1 = 1$:

$$\begin{aligned} a_n &= \frac{n-1}{n}a_{n-1}, \quad n \geq 1 \\ &= 0, \quad \forall n \geq 1 \end{aligned}$$

Let $r = r_2 = \dfrac{1}{2}$:

$$\begin{aligned} b_n &= \frac{n-\frac{3}{2}}{n-\frac{1}{2}}b_{n-1} \\ &= \frac{2n-3}{2n-1}b_{n-1}, \quad n \geq 1 \end{aligned}$$

From this relation we obtain

$$\begin{aligned} b_1 &= \frac{-1}{1}b_0, \quad b_2 = \frac{1}{3}b_1 = \frac{-1\cdot 1}{1\cdot 3}b_0 \\ b_3 &= \frac{3}{5}b_2 = \frac{-1\cdot 1\cdot 3}{1\cdot 3\cdot 5}b_0 \\ &\vdots \\ b_n &= \frac{-1\cdot 1\cdot 3\cdot 5\cdot \ldots \cdot (2n-3)}{1\cdot 3\cdot 5\cdot 7\cdot \ldots \cdot (2n-1)}b_0 \\ &= \frac{-b_0}{2n-1}, \quad n \geq 1 \end{aligned}$$

Hence,

$$y_1(x) = \sum_{n=0}^{\infty} a_n x^{n+1}$$

$$\begin{aligned} &= a_0 x \\ y_2(x) &= \sum_{n=0}^{\infty} b_n x^{n+\frac{1}{2}} \\ &= b_0 x^{\frac{1}{2}} - \sum_{n=1}^{\infty} \frac{x^{n+\frac{1}{2}}}{2n-1} b_0 \end{aligned}$$

both valid, at least, in the region $0 < x < 1$. Consequently, the general solution of (F2.1) 'near $x = 0$' is given by

$$y(x) = a_0 x + b_0 \left[x^{\frac{1}{2}} - \sum_{n=1}^{\infty} \frac{x^{n+\frac{1}{2}}}{2n-1} \right], \qquad 0 < x < 1 \text{ (at least)}$$

where a_0 and b_0 are arbitrary (non-zero) constants.

FINAL EXAMINATION #3

SOLUTIONS

In what follows, c_i, $i = 1, 2, ...$ will denote arbitrary constants. Rules of logarithms and exponentials, a summary of the main techniques of integration as well as a *table of integrals* can be found in the Appendix.

Question 1.(15%) Solve the boundary value problem

$$\frac{d^2y}{dt^2}\cos\left(\frac{dy}{dt}\right) = \cos t, \qquad -\frac{\pi}{2} \leq t \leq \frac{\pi}{2}, \qquad y(1) = 1, \; \frac{dy}{dt}(0) = 0$$

Comment on the case $t \notin \left[-\frac{\pi}{2}, \frac{\pi}{2}\right]$.

Solution. The differential equation is of the type in which the dependent variable y does not appear explicitly (see Section 3.7). We make the substitution

$$\frac{dy}{dt} = p, \quad \frac{d^2y}{dt^2} = \frac{dp}{dt}$$

and obtain

$$\frac{dp}{dt}\cos p = \cos t$$

This equation is now separable (Section 2.1).

$$\begin{aligned} \int \cos p dp &= \int \cos t dt \\ \sin p &= \sin t + c_1 \end{aligned}$$

Apply the derivative boundary condition:

$$p(0) = \frac{dy}{dt}(0) = 0$$

to obtain

$$\begin{aligned} \sin 0 &= \sin 0 + c_1 \\ 0 &= c_1 \end{aligned}$$

Hence, we have that

$$\sin p = \sin t$$

Since $-\frac{\pi}{2} \leq t \leq \frac{\pi}{2}$,we can write

$$\begin{aligned} p &= \arcsin(\sin t) \\ &= t \end{aligned}$$

Thus,

$$\begin{aligned} p &= \frac{dy}{dt} = t \\ y(t) &= \frac{t^2}{2} + c_2 \end{aligned}$$

Applying the boundary condition $y(1) = 1$, we obtain

$$\begin{aligned} 1 &= \frac{1}{2} + c_2 \\ c_2 &= \frac{1}{2} \end{aligned}$$

The solution of the boundary value problem is given by

$$y(t) = \frac{t^2}{2} + \frac{1}{2}, \quad -\frac{\pi}{2} \le t \le \frac{\pi}{2}$$

If $t \notin \left[-\frac{\pi}{2}, \frac{\pi}{2}\right]$, we cannot say that $\arcsin(\sin t) = t$ since this is true only when $t \in \left[-\frac{\pi}{2}, \frac{\pi}{2}\right]$. This means that we have the following nonlinear equation relating $p = \frac{dy}{dt}$ to t.

$$\sin p = \sin t$$

Question 2. (10%) Classify all singular points (*excluding* the point at infinity) of the following ordinary differential equation

$$\left(1 - x^2\right)^2 x \frac{d^2y}{dx^2} - 2x(1 + x)\frac{dy}{dx} + \beta(\beta - 1)y = 0$$

Here, β is a real number.

Solution. Write the differential equation in the form

$$\frac{d^2y}{dx^2} - \frac{2x(1 + x)}{\left(1 - x^2\right)^2 x}\frac{dy}{dx} + \frac{\beta(\beta - 1)}{\left(1 - x^2\right)^2 x} y = 0$$

As in Section 3.8, we identify

$$\begin{aligned} P(x) &= -\frac{2x(1 + x)}{x\left(1 - x^2\right)^2} \\ Q(x) &= \frac{\beta(\beta - 1)}{\left(1 - x^2\right)^2 x} \end{aligned}$$

Since both $P(x)$ and $Q(x)$ fail to be analytic at $x = 0, \pm 1$, we have singular points at $x = 0, \pm 1$. We classify each of these singular points as follows (see Section 3.8).

$x = 0$:

$$\begin{aligned}
p(x) &= xP(x) \\
&= -\frac{2x^2(1+x)}{x(1-x^2)^2} \\
&= -\frac{2x(1+x)}{(1-x^2)^2}
\end{aligned}$$

$$\begin{aligned}
q(x) &= x^2 Q(x) \\
&= x^2 \frac{\beta(\beta-1)}{(1-x^2)^2 x} \\
&= \frac{x\beta(\beta-1)}{(1-x^2)^2}
\end{aligned}$$

Since both p and q are analytic at $x = 0$, the latter is a *regular singular point.*

$x = 1$:

$$\begin{aligned}
p(x) &= (x-1)P(x) \\
&= -\frac{2x(1+x)(x-1)}{x(1-x^2)^2} \\
&= -\frac{2}{(x-1)(1+x)}
\end{aligned}$$

$$\begin{aligned}
q(x) &= (x-1)^2 Q(x) \\
&= (x-1)^2 \frac{\beta(\beta-1)}{(1-x^2)^2 x} \\
&= \frac{\beta(\beta-1)}{(1+x)^2 x}
\end{aligned}$$

Since p is not analytic at $x = 1$, the latter is an *irregular singular point.*

$x = -1:$

$$\begin{aligned} p(x) &= (x+1)\,P(x) \\ &= -\frac{2x(1+x)^2}{x(1-x^2)^2} \\ &= -\frac{2}{(x-1)^2} \end{aligned}$$

$$\begin{aligned} q(x) &= (x+1)^2\,Q(x) \\ &= (x+1)^2\,\frac{\beta(\beta-1)}{(1-x^2)^2\,x} \\ &= \frac{\beta(\beta-1)}{(x-1)^2\,x} \end{aligned}$$

Since p and q are both analytic at $x = -1$, the latter is an *regular singular point.*

3. **(i)** (10%) Given that $y = x$ is one solution of the homogeneous equation

$$\left(x^2+1\right)y'' - 2xy' + 2y = 0$$

find its general solution.
(ii) (15%) Find the general solution of the inhomogeneous equation

$$\frac{d^2y}{dx^2} - 9\frac{dy}{dx} + 18y = \sin\left(e^{-3x}\right)$$

Solution

Question 3.(i) We are given one solution of the homogeneous equation. This suggests that we use *Reduction of Order* (see Section 3.4) to find the other linearly independent solution.

$$\begin{aligned} y &= vx \\ y' &= v + xv' \\ y'' &= xv'' + 2v' \end{aligned}$$

Substitute into the homogeneous differential equation

$$\begin{aligned} \left(x^2+1\right)y'' - 2xy' + 2y &= 0 \\ \left(x^2+1\right)\left(xv''+2v'\right) - 2x\left(v+xv'\right) + 2xv &= 0 \\ x\left(x^2+1\right)v'' + 2v' &= 0 \\ v'' + \frac{2}{x\left(x^2+1\right)}v' &= 0 \end{aligned}$$

Let $w = v'$.

$$w' + \frac{2}{x\left(x^2+1\right)}w = 0$$

This equation is first order in w (see Section 2.5). An integrating factor is given by

$$\begin{aligned}
\mu(x) &= \exp\left(2\int \underbrace{\frac{dx}{x\left(x^2+1\right)}}_{\text{Partial Fractions}}\right) \\
&= \exp\left(2\int\left(\frac{1}{x} - \underbrace{\frac{x}{x^2+1}}_{\text{Let } u=x^2+1}\right)dx\right) \\
&= \exp\left(2\left[\ln|x| - \frac{1}{2}\ln\left|x^2+1\right|\right]\right) \\
&= \exp\left(\ln\frac{x^2}{x^2+1}\right) \\
&= \frac{x^2}{x^2+1}
\end{aligned}$$

The differential equation now becomes

$$\frac{d}{dx}\left(w\frac{x^2}{x^2+1}\right) = 0$$

Integrate both sides with respect to x.

$$\begin{aligned}
\frac{x^2}{x^2+1}w &= c_1 \\
w &= v' = \frac{c_1\left(x^2+1\right)}{x^2} \\
&= c_1 + \frac{c_1}{x^2}
\end{aligned}$$

Integrate again with respect to x.

$$v = c_1 x - \frac{c_1}{x} + c_2$$

Finally, the general solution of the homogeneous differential equation is given by

$$\begin{aligned}
y &= vx \\
&= c_1 x^2 - c_1 + c_2 x \\
&= c_1\left(x^2-1\right) + c_2 x
\end{aligned}$$

Question 3.(ii) The equation

$$\frac{d^2y}{dx^2} - 9\frac{dy}{dx} + 18y = \sin\left(e^{-3x}\right)$$

is linear but the right-hand side is *non-standard* in that it is not accommodated by Table 3.3.1. However, we can still write the general solution in the form

$$y(x) = y_c(x) + y_p(x)$$

For y_c, the characteristic equation is given by

$$\begin{aligned} m^2 - 9m + 18 &= 0 \\ (m-6)(m-3) &= 0 \\ m &= 3, 6 \end{aligned}$$

Hence,

$$y_c(x) = c_1 e^{3x} + c_2 e^{6x}$$

To find a particular solution $y_p(x)$, we use *Variation of Parameters* (Section 3.5) (setting all constants of integration to zero). Let

$$y_p(x) = A(x)\, e^{3x} + B(x)\, e^{6x}$$

where A and B are functions of x given by the matrix equation

$$\begin{bmatrix} e^{3x} & e^{6x} \\ 3e^{3x} & 6e^{6x} \end{bmatrix} \begin{bmatrix} A'(x) \\ B'(x) \end{bmatrix} = \begin{bmatrix} 0 \\ \sin\left(e^{-3x}\right) \end{bmatrix}$$

Solving this matrix equation (by multiplying both sides by the inverse of the matrix $\begin{bmatrix} e^{3x} & e^{6x} \\ 3e^{3x} & 6e^{6x} \end{bmatrix}$), we obtain

$$\begin{aligned} \begin{bmatrix} A'(x) \\ B'(x) \end{bmatrix} &= \begin{bmatrix} 2e^{-3x} & -\frac{1}{3}e^{-3x} \\ -e^{-6x} & \frac{1}{3}e^{-6x} \end{bmatrix} \begin{bmatrix} 0 \\ \sin\left(e^{-3x}\right) \end{bmatrix} \\ &= \begin{bmatrix} -\frac{1}{3}e^{-3x}\sin\left(e^{-3x}\right) \\ \frac{1}{3}e^{-6x}\sin\left(e^{-3x}\right) \end{bmatrix} \end{aligned}$$

Hence,

$$\begin{aligned} A(x) &= -\frac{1}{3}\int e^{-3x}\sin\left(e^{-3x}\right) dx \\ B(x) &= \frac{1}{3}\int e^{-6x}\sin\left(e^{-3x}\right) dx \end{aligned}$$

To determine each of these integrals, we make the substitution $u = e^{-3x}$, so that $du = -3e^{-3x}dx = -3udx$. Then,

$$\begin{aligned} A(x) &= \frac{1}{9}\int \sin u du \\ &= -\frac{1}{9}\cos u \\ &= -\frac{1}{9}\cos\left(e^{-3x}\right) \end{aligned}$$

Similarly, using integration by parts,

$$\begin{aligned} B(x) &= -\frac{1}{9}\int u \sin u du \\ &= -\frac{1}{9}(\sin u - u\cos u) \\ &= -\frac{1}{9}\left(\sin\left(e^{-3x}\right) - e^{-3x}\cos\left(e^{-3x}\right)\right) \end{aligned}$$

Finally,

$$\begin{aligned} y_p(x) &= A(x)e^{3x} + B(x)e^{6x} \\ &= -\frac{1}{9}e^{3x}\cos\left(e^{-3x}\right) + e^{6x}\left(-\frac{1}{9}\left(\sin\left(e^{-3x}\right) - e^{-3x}\cos\left(e^{-3x}\right)\right)\right) \\ &= -\frac{1}{9}e^{6x}\sin\left(e^{-3x}\right) \end{aligned}$$

The general solution of the differential equation is therefore given by

$$\begin{aligned} y(x) &= y_c(x) + y_p(x) \\ &= c_1e^{3x} + c_2e^{6x} - \frac{1}{9}e^{6x}\sin\left(e^{-3x}\right) \end{aligned}$$

Question 4.(10%) Use the method of Laplace transforms to solve the following initial value problem.

$$\begin{aligned} \frac{dy}{dt} + x(t) &= A(t) \\ \frac{dx}{dt} - y(t) &= B(t) \\ x(0) &= 0, \quad y(0) = 1 \end{aligned}$$

where A and B are given functions of t. (You may leave your answer in terms of integrals).

Solution. First take the Laplace transform of the system (see Example 3.9.5).

$$\begin{aligned} L[y'(t)] + L[x(t)] &= L[A(t)] \\ L[x'(t)] - L[y(t)] &= L[B(t)] \end{aligned}$$

From Table 3.9.1, we obtain

$$\begin{aligned} sY(s) - y(0) + X(s) &= L[A(t)] \\ sX(s) - x(0) - Y(s) &= L[B(t)] \end{aligned}$$

where $X(s) = L[x(t)]$ and $Y(s) = L[y(t)]$. Applying the given initial conditions, we obtain

$$\begin{aligned} sY(s) + X(s) &= L[A(t)] + 1 \\ sX(s) - Y(s) &= L[B(t)] \end{aligned}$$

Solving this system for $X(s)$ and $Y(s)$ (by elimination, Cramer's rule or matrix methods), we obtain

$$\begin{aligned} Y(s) &= \frac{-L[B(t)] + sL[A(t)] + s}{s^2+1} \\ X(s) &= \frac{L[A(t)] + 1 + sL[B(t)]}{s^2+1} \end{aligned}$$

Take the inverse Laplace transform:

$$\begin{aligned} x(t) &= L^{-1}\left[\frac{1}{s^2+1}\right] + L^{-1}\left[\frac{s}{s^2+1}L[B(t)]\right] + L^{-1}\left[\frac{1}{s^2+1}L[A(t)]\right] \\ y(t) &= L^{-1}\left[\frac{s}{s^2+1}\right] + L^{-1}\left[\frac{s}{s^2+1}L[A(t)]\right] - L^{-1}\left[\frac{1}{s^2+1}L[B(t)]\right] \end{aligned}$$

From Tables 3.9.1 and 3.9.2, we have

$$\begin{aligned} x(t) &= \sin t + \int_0^t f_1(t)\, B(t-\beta)\, d\beta + \int_0^t f_2(t) A(t-\beta) d\beta \\ y(t) &= \cos t + \int_0^t f_1(t)\, A(t-\beta)\, d\beta - \int_0^t f_2(t) B(t-\beta) d\beta \end{aligned}$$

where

$$\begin{aligned} f_1(t) &= L^{-1}\left[\frac{s}{s^2+1}\right] = \cos t \\ f_2(t) &= L^{-1}\left[\frac{1}{s^2+1}\right] = \sin t \end{aligned}$$

Finally, the solution of the initial value problem is given by

$$\begin{aligned} x(t) &= \sin t + \int_0^t \cos t B(t-\beta)\, d\beta + \int_0^t \sin t A(t-\beta) d\beta \\ y(t) &= \cos t + \int_0^t \cos t A(t-\beta)\, d\beta - \int_0^t \sin t B(t-\beta) d\beta \end{aligned}$$

5. (i)(20%) Find the general solution in series about $x = 0$.

$$x\frac{d^2y}{dx^2} + 2\frac{dy}{dx} + 11xy = 0, \qquad x > 0$$

(ii)(5%) Write the general solution from 5(i) above in terms of known functions.

Solution

Question 5.(i) From Section 3.8, $x = 0$ is an *ordinary* point of the differential equation if the expressions $P(x)$ and $Q(x)$ in (3.8.4) of Chapter 3 are *both* analytic at $x = 0$. In fact, here

$$\begin{aligned} P(x) &= \frac{2}{x} \\ Q(x) &= \frac{11x}{x} \end{aligned}$$

It is clear that neither $P(x)$ nor $Q(x)$ are analytic at $x = 0$ so that $x = 0$ is a *singular* point of the differential equation. According to the theory in Section 3.8, $x = 0$ is a *regular singular point* if $p(x) = xP(x)$ and $q(x) = x^2Q(x)$ are *both* analytic at $x = 0$. In fact, here,

$$\begin{aligned} p(x) &= xP(x) = 2 \\ q(x) &= x^2Q(x) = 11x^2 \end{aligned}$$

which are indeed *both* analytic at $x = 0$. Hence, $x = 0$ is a *regular singular point* and, by Theorem 3.8.6, for $x > 0$, the differential equation has at least one Frobenius series solution (corresponding to the larger root of the indicial equation (3.8.13)). To see which form a second linearly independent solution will take, we examine the roots of the indicial equation (3.8.13) .

$$\begin{aligned} r(r-1) + p(0)r + q(0) &= 0 \\ r(r-1) + 2r &= 0 \\ r(r+1) &= 0 \\ r_1 &= 0, \quad r_2 = -1 \end{aligned}$$

The roots of the indicial equation are such that $r_1 - r_2$ is a positive *integer*. Hence, by Theorem 3.8.6(iii), the second linearly independent solution may or may not be a Frobenius series (depending on whether or not the logarithmic term in Theorem 3.8.6(iii) vanishes, that is, whether or not $C = 0$ in Theorem 3.8.6(iii)). To see which is the case, we start with the smaller root r_2 of the indicial equation and obtain a recurrence relation. If a second linearly independent Frobenius solution exists, this procedure will yield *both* Frobenius series solutions (those corresponding to r_2 *and* r_1) *at the same time.* If it doesn't, we know that the second linearly independent solution is not a Frobenius series and does, in fact, involve a logarithmic term. Hence, substitute the Frobenius series (corresponding to the smaller root $r_2 = -1$)

$$y(x) = \sum_{n=0}^{\infty} a_n x^{n-1} \quad , \qquad a_0 \neq 0$$

into the differential equation

$$x\frac{d^2y}{dx^2} + 2\frac{dy}{dx} + 11xy = 0$$

This gives

$$\sum_{n=0}^{\infty} a_n (n-1)(n-2)x^{n-2} + 2\sum_{n=0}^{\infty} a_n (n-1) x^{n-2} + 11\sum_{n=0}^{\infty} a_n x^n \ = 0$$

Simplifying and shifting summation index in the third sum ($n \to n-2$), we obtain

$$\sum_{n=0}^{\infty} a_n n (n-1) x^{n-2} + 11\sum_{n=2}^{\infty} a_{n-2} x^{n-2} = 0$$

Equate coefficients of x^n for different values of n :

$$\begin{aligned}
n &= 0: \ a_0 \cdot 0 \cdot (-1) = 0, \qquad a_0 \neq 0 \quad \therefore \ a_0 = \text{arbitrary} \\
n &= 1: \ a_1 \cdot 1 \cdot 0 = 0 \quad \therefore \ a_1 = \text{arbitrary} \\
n &\geq 2: \ a_n = -\frac{11}{n(n-1)} a_{n-2}
\end{aligned}$$

Hence, we have *two* arbitrary constants a_0 and a_1 so that we expect to obtain *two* linearly independent Frobenius series solutions (the existence of the second linearly independent Frobenius series solution depends entirely on the existence of *two* arbitrary constants at this stage of the procedure - if only one arbitrary constant emerges at this stage, no second linearly independent Frobenius series solution exists. Instead, Theorem 3.8.6 (iii) guarantees a logarithmic solution). From the recurrence relation for $n \geq 2$, we obtain

$$\begin{aligned}
a_0 &= a_0 & a_1 &= a_1 \\
a_2 &= -\tfrac{11a_0}{2\cdot 1} & a_3 &= -\tfrac{11}{3\cdot 2} a_1 \\
a_4 &= -\tfrac{11a_2}{4\cdot 3} = \tfrac{11^2 a_0}{4\cdot 3\cdot 2\cdot 1} & a_5 &= -\tfrac{11}{5\cdot 4} a_3 = \tfrac{11^2}{5\cdot 4\cdot 3\cdot 2} a_1 \\
a_6 &= -\tfrac{11a_4}{6\cdot 5} = -\tfrac{11^3 a_0}{6!} & a_7 &= -\tfrac{11}{7\cdot 6} a_5 = -\tfrac{11^3}{7!} a_1 \\
&\vdots \\
&\vdots \\
a_{2n} &= \tfrac{(-1)^n 11^n}{(2n)!} a_0 \ , \quad n \geq 0 & a_{2n+1} &= \tfrac{(-1)^n 11^n}{(2n+1)!} a_1 \ , \quad n \geq 0
\end{aligned}$$

Hence, our general solution is given by

$$\begin{aligned}
y(x) &= \sum_{n=0}^{\infty} a_n x^{n-1} \\
&= x^{-1} \sum_{n=0}^{\infty} a_n x^n \\
&= x^{-1} \left[\sum_{n=0}^{\infty} a_{2n} x^{2n} + \sum_{n=0}^{\infty} a_{2n+1} x^{2n+1} \right] \\
&= x^{-1} \left[\sum_{n=0}^{\infty} \frac{(-1)^n 11^n a_0}{(2n)!} x^{2n} + \sum_{n=0}^{\infty} \frac{(-1)^n 11^n a_1}{(2n+1)!} x^{2n+1} \right] \\
&= a_0 \sum_{n=0}^{\infty} \frac{(-1)^n 11^n}{(2n)!} x^{2n-1} + a_1 \sum_{n=0}^{\infty} \frac{(-1)^n 11^n}{(2n+1)!} x^{2n}
\end{aligned}$$

Question 5.(ii) The general solution

$$y(x) = a_0 \sum_{n=0}^{\infty} \frac{(-1)^n 11^n}{(2n)!} x^{2n-1} + a_1 \sum_{n=0}^{\infty} \frac{(-1)^n 11^n}{(2n+1)!} x^{2n}$$

can also be written as

$$\begin{aligned}
y(x) &= x^{-1} \left[a_0 \sum_{n=0}^{\infty} \frac{(-1)^n \left(\sqrt{11}x\right)^{2n}}{(2n)!} + \frac{a_1}{\sqrt{11}} \sum_{n=0}^{\infty} \frac{(-1)^n \left(\sqrt{11}x\right)^{2n+1}}{(2n+1)!} \right] \\
&= x^{-1} \left[a_0 \sum_{n=0}^{\infty} \frac{(-1)^n \left(\sqrt{11}x\right)^{2n}}{(2n)!} + c_1 \sum_{n=0}^{\infty} \frac{(-1)^n \left(\sqrt{11}x\right)^{2n+1}}{(2n+1)!} \right] \\
&= x^{-1} \left(a_0 \cos\left(\sqrt{11}x\right) + c_1 \sin\left(\sqrt{11}x\right) \right)
\end{aligned}$$

where, a_0 and c_1 $\left(= \frac{a_1}{\sqrt{11}}\right)$ are arbitrary (nonzero) constants.

Question 6.(15%) Find the general solution of the following system of equations.

$$\begin{aligned}
\frac{dx}{dt} &= 3x + 2y + 2z \\
\frac{dy}{dt} &= -5x - 4y - 2z \\
\frac{dz}{dt} &= 5x + 5y + 3z
\end{aligned}$$

Solution. Write the system in matrix form and use the theory of eigenvalues and eigenvectors in Section 3.10.

$$\begin{bmatrix} x' \\ y' \\ z' \end{bmatrix} = \begin{bmatrix} 3 & 2 & 2 \\ -5 & -4 & -2 \\ 5 & 5 & 3 \end{bmatrix} \begin{bmatrix} x \\ y \\ z \end{bmatrix}$$

The eigenvalues of the matrix

$$\begin{bmatrix} 3 & 2 & 2 \\ -5 & -4 & -2 \\ 5 & 5 & 3 \end{bmatrix}$$

are given by the determinant equation

$$\begin{aligned} \begin{vmatrix} 3-\lambda & 2 & 2 \\ -5 & -4-\lambda & -2 \\ 5 & 5 & 3-\lambda \end{vmatrix} &= 0 \\ \left(\lambda^2+\lambda-2\right)(3-\lambda) &= 0 \\ (3-\lambda)(2+\lambda)(\lambda-1) &= 0 \\ \lambda &= -2, 1, 3 \end{aligned}$$

Hence, the eigenvalues are given by

$$\lambda_1 = -2, \lambda_2 = 1, \lambda_3 = 3$$

which are *real and distinct.* Next, find the corresponding eigenvectors.

$\lambda_1 = -2:$

Let the corresponding eigenvector be denoted by $\mathbf{v}_1 = \begin{bmatrix} c_1 & c_2 & c_3 \end{bmatrix}^T$ where the c_i, $i = 1, 2, 3$ are constant. Then, $\mathbf{v}_1$ is given by

$$\begin{bmatrix} 5 & 2 & 2 \\ -5 & -2 & -2 \\ 5 & 5 & 5 \end{bmatrix} \begin{bmatrix} c_1 \\ c_2 \\ c_3 \end{bmatrix} = \begin{bmatrix} 0 \\ 0 \\ 0 \end{bmatrix}$$

This is equivalent to the system

$$\begin{aligned} 5c_1 + 2c_2 + 2c_3 &= 0 \\ c_1 + c_2 + c_3 &= 0 \end{aligned}$$

which has solution

$$\begin{aligned} c_1 &= 0 \\ c_2 &= \text{arbitrary} \\ c_3 &= -c_2 \end{aligned}$$

Hence, the corresponding eigenvector is given by

$$\mathbf{v}_1 = \begin{bmatrix} 0 \\ 1 \\ -1 \end{bmatrix}$$

$\lambda_1 = 1:$

Let the corresponding eigenvector be denoted by $\mathbf{v}_2 = \begin{bmatrix} c_1 & c_2 & c_3 \end{bmatrix}^T$ where the c_i, $i = 1, 2, 3$ are constant. Then, $\mathbf{v}_2$ is given by

$$\begin{bmatrix} 2 & 2 & 2 \\ -5 & -5 & -2 \\ 5 & 5 & 2 \end{bmatrix} \begin{bmatrix} c_1 \\ c_2 \\ c_3 \end{bmatrix} = \begin{bmatrix} 0 \\ 0 \\ 0 \end{bmatrix}$$

This is equivalent to the system

$$\begin{aligned} c_1 + c_2 + c_3 &= 0 \\ 5c_1 + 5c_2 + 2c_3 &= 0 \end{aligned}$$

which has solution

$$\begin{aligned} c_3 &= 0 \\ c_1 &= \text{arbitrary} \\ c_2 &= -c_1 \end{aligned}$$

Hence, the corresponding eigenvector is given by

$$\mathbf{v}_2 = \begin{bmatrix} 1 \\ -1 \\ 0 \end{bmatrix}$$

$\lambda_1 = 3:$

Let the corresponding eigenvector be denoted by $\mathbf{v}_3 = \begin{bmatrix} c_1 & c_2 & c_3 \end{bmatrix}^T$ where the c_i, $i = 1, 2, 3$ are constant. Then, $\mathbf{v}_3$ is given by

$$\begin{bmatrix} 0 & 2 & 2 \\ -5 & -7 & -2 \\ 5 & 5 & 0 \end{bmatrix} \begin{bmatrix} c_1 \\ c_2 \\ c_3 \end{bmatrix} = \begin{bmatrix} 0 \\ 0 \\ 0 \end{bmatrix}$$

This is equivalent to the system

$$\begin{aligned} c_2 + c_3 &= 0 \\ c_1 + c_2 &= 0 \end{aligned}$$

which has solution

$$\begin{aligned} c_1 &= -c_2 = c_3 \\ c_2 &= -c_3 \\ c_3 &= \text{arbitrary} \end{aligned}$$

Hence, the corresponding eigenvector is given by

$$\mathbf{v}_3 = \begin{bmatrix} 1 \\ -1 \\ 1 \end{bmatrix}$$

The general solution of the system is now given by (see (3.10.7))

$$\begin{aligned} \mathbf{x}(t) &= \begin{bmatrix} x \\ y \\ z \end{bmatrix} \\ &= c_1\mathbf{v}_1 e^{\lambda_1 t} + c_2\mathbf{v}_2 e^{\lambda_2 t} + c_3\mathbf{v}_3 e^{\lambda_3 t} \\ &= c_1 \begin{bmatrix} 0 \\ 1 \\ -1 \end{bmatrix} e^{-2t} + c_2 \begin{bmatrix} 1 \\ -1 \\ 0 \end{bmatrix} e^{t} + c_3 \begin{bmatrix} 1 \\ -1 \\ 1 \end{bmatrix} e^{3t} \end{aligned}$$

or

$$\begin{aligned} x(t) &= c_2 e^{t} + c_3 e^{3t} \\ y(t) &= c_1 e^{-2t} - c_2 e^{t} - c_3 e^{3t} \\ z(t) &= -c_1 e^{-2t} + c_3 e^{3t} \end{aligned}$$

FINAL EXAMINATION #4
SOLUTIONS

In what follows, c_i, $i = 1, 2, ...$ will denote arbitrary constants. Rules of logarithms and exponentials, a summary of the main techniques of integration as well as a *table of integrals* can be found in the Appendix.

Question 1.(10%) Solve the initial value problem

$$\frac{dy}{dx} = -\frac{(x+1)\tan y}{x\sec^2 y}, \qquad y(1) = \frac{\pi}{4}$$

Solution. The differential equation is neither separable, exact, homogeneous, linear nor Bernouilli. We try to find an integrating factor (see Section 2.4). We first write the differential equation in the standard form (2.3.1).

$$\underbrace{(x+1)\tan y}_{M}\, dx + \underbrace{x\sec^2 y}_{N}\, dy = 0$$

Next, we determine the required partial derivatives with each of (2.4.1) and (2.4.3) in mind.

$$\frac{\partial M}{\partial y} = (x+1)\sec^2 y; \qquad \frac{\partial N}{\partial x} = \sec^2 y$$

Examine (2.4.1)

$$\begin{aligned} \frac{1}{N}\left(\frac{\partial M}{\partial y} - \frac{\partial N}{\partial x}\right) &= \frac{1}{x\sec^2 y}\left(x\sec^2 y\right) \\ &= 1 \\ &= R(x) \end{aligned}$$

- a function of x (or y) only! Hence, an integrating factor $\mu(x)$, say, can be found from (2.4.2).

$$\begin{aligned} \mu(x) &= \exp(\int 1dx) \\ &= \exp(x) \\ &= e^x + c_1 \end{aligned}$$

Choose

$$\mu(x) = e^x$$

Next, we multiply both sides of our differential equation by $\mu(x) = e^x$.

$$\underbrace{e^x (x+1) \tan y \, dx}_{M} + \underbrace{e^x x \sec^2 y}_{N} \, dy = 0$$

This equation is now exact (in the sense of Section 2.3) since

$$\frac{\partial M}{\partial y} = e^x (x+1) \sec^2 y = \frac{\partial N}{\partial x}$$

Its general solution can be now be found as in Section 2.3 as follows. The resulting exact equation is given by

$$\underbrace{e^x (x+1) \tan y \, dx}_{M} + \underbrace{e^x x \sec^2 y}_{N} \, dy = 0$$

Hence, there exists a function $F(x, y)$ such that

$$\frac{\partial F}{\partial x} = M = e^x (x+1) \tan y \quad \text{(F4.1)}$$

$$\frac{\partial F}{\partial y} = N = e^x x \sec^2 y \quad \text{(F4.2)}$$

From (F4.2)

$$F(x, y) = e^x x \tan y \ + f(x) \quad \text{(F4.3)}$$

where f is arbitrary. To find f we require that (F4.3) satisfies (F4.1):

$$e^x (x+1) \tan y + f'(x) = e^x (x+1) \tan y$$

It follows that

$$\begin{aligned} f'(x) &= 0 \\ f(x) &= c_2 \end{aligned}$$

Hence,

$$F(x, y) = e^x x \tan y \ + c_2$$

and, the general solution is given by

$$\begin{aligned} F(x, y) &= c_3 \\ e^x x \tan y \ + c_2 &= c_3 \\ e^x x \tan y &= c_4 \end{aligned}$$

where $c_4 = c_3 - c_2$. Finally, applying the given initial condition $y(1) = \frac{\pi}{4}$, we obtain

$$\begin{aligned} e(1)(1) &= c_4 \\ c_4 &= e \end{aligned}$$

The solution of the initial value problem is given by

$$e^x x \tan y = e$$

Question 2.(15%) Use the method of Laplace transforms to solve the following initial value problem.

$$\frac{d^2y}{dt^2} + y(t) = H(t), \qquad y(0) = 1, \; \frac{dy}{dt}(0) = 1$$

where $H(t)$ is given by

$$H(t) = \begin{cases} t, & 0 \leq t < 1, \\ 0, & t \geq 1 \end{cases}$$

Solution. First take the Laplace transform of both sides of the differential equation. From Table 3.9.2,

$$\begin{aligned} L[\frac{d^2y}{dt^2} + y(t)] &= L[H(t)] \\ s^2 L[y] - sy(0) - \frac{dy}{dt}(0) + L[y] &= L[H(t)] \end{aligned}$$

Applying the given initial conditions, we obtain

$$\begin{aligned} s^2 L[y] - s - 1 + L[y] &= L[H(t)] \\ \left(s^2 + 1\right) L[y] &= s + 1 + L[H(t)] \end{aligned} \tag{F4.4}$$

To find the Laplace transform of the function $H(t)$, we first write it in terms of the Heaviside function $\alpha(t-c)$ (see Table 3.9.2).

$$\begin{aligned} H(t) &= t + \begin{cases} 0, & 0 \leq t < 1, \\ -t, & t \geq 1 \end{cases} \\ &= t - t \begin{cases} 0, & 0 \leq t < 1, \\ 1, & t \geq 1 \end{cases} \\ &= t\left[1 - \alpha(t-1)\right] \end{aligned}$$

Hence, from Table 3.9.1,

$$\begin{aligned} L[H(t)] &= L[t] - \underbrace{L[t\alpha(t-1)]}_{\text{Table 3.9.2}} \\ &= \frac{1}{s^2} - e^{-s} L[F(t)] \end{aligned}$$

where $F(t-1) = t$ so that $F(t) = t+1$. Hence,

$$\begin{aligned} L[H(t)] &= \frac{1}{s^2} - e^{-s} L[F(t)] \\ &= \frac{1}{s^2} - e^{-s} L[t+1] \\ &= \frac{1}{s^2} - e^{-s}\left(\frac{1}{s^2} + \frac{1}{s}\right) \end{aligned}$$

From (F4.4),

$$\begin{aligned} \left(s^2+1\right) L[y] &= s+1+\frac{1}{s^2} - e^{-s}\left(\frac{1}{s^2}+\frac{1}{s}\right) \\ L[y] &= \frac{s}{s^2+1} + \frac{1}{s^2+1} + \frac{1}{s^2\left(s^2+1\right)} - \frac{e^{-s}}{s^2\left(s^2+1\right)} - \frac{e^{-s}}{s\left(s^2+1\right)} \end{aligned}$$

Inverting the Laplace transform and using Tables 3.9.1 and 3.9.2, we have

$$\begin{aligned} y(t) &= L^{-1}[\frac{s}{s^2+1}] + L^{-1}[\frac{1}{s^2+1}] + L^{-1}[\frac{1}{s^2\left(s^2+1\right)}] - L^{-1}[\frac{e^{-s}}{s^2\left(s^2+1\right)}] - L^{-1}[\frac{e^{-s}}{s\left(s^2+1\right)}] \\ &= \cos t + \sin t + F_1(t) - \alpha(t-1) F_1(t-1) - \alpha(t-1) F_2(t-1) \qquad \text{(F4.5)} \end{aligned}$$

where

$$\begin{aligned} F_1(t) &= L^{-1}[\frac{1}{s^2\left(s^2+1\right)}] \\ F_2(t) &= L^{-1}[\frac{1}{s\left(s^2+1\right)}] \end{aligned}$$

Each of these inverse transforms can be found using partial fractions (see Appendix) and Table 3.9.1:

$$\begin{aligned} F_1(t) &= L^{-1}[\frac{1}{s^2\left(s^2+1\right)}] \\ &= L^{-1}[\frac{1}{s^2} - \frac{1}{s^2+1}] \\ &= t - \sin t \end{aligned}$$

Hence,

$$F_1(t-1) = (t-1) - \sin(t-1)$$

Similarly,

$$\begin{aligned} F_2(t) &= L^{-1}[\frac{1}{s\left(s^2+1\right)}] \\ &= L^{-1}[\frac{1}{s} - \frac{s}{s^2+1}] \\ &= 1 - \cos t \end{aligned}$$

Hence,

$$F_2(t-1) = 1 - \cos(t-1)$$

Finally, from (F4.5),

$$\begin{aligned} y(t) &= \cos t + \sin t + F_1(t) - \alpha(t-1) F_1(t-1) - \alpha(t-1) F_2(t-1) \\ &= \cos t + \sin t + t - \sin t - \alpha(t-1)[(t-1) - \sin(t-1)] - \alpha(t-1)[1 - \cos(t-1)] \\ &= \cos t + t - \alpha(t-1)[t - \sin(t-1) - \cos(t-1)] \end{aligned}$$

Question 3.(10%) Find the general solution of the equation

$$(e^x + 1)\frac{d^2y}{dx^2} - (e^x + 1)\frac{dy}{dx} = 1$$

Solution. Since $e^x + 1 \neq 0$, we can write the equation in the form

$$\frac{d^2y}{dx^2} - \frac{dy}{dx} = \frac{1}{e^x + 1}$$

The equation is linear but the right-hand side is *non-standard* in that it is not accommodated by Table 3.3.1. However, we can still write the general solution in the form

$$y(x) = y_c(x) + y_p(x)$$

For y_c, the characteristic equation is given by

$$\begin{aligned} m^2 - m &= 0 \\ m(m-1) &= 0 \\ m &= 0, 1 \end{aligned}$$

Hence,

$$y_c(x) = c_1 + c_2 e^x$$

To find a particular solution $y_p(x)$, we use *Variation of Parameters* (Section 3.5) (setting all constants of integration to zero). Let

$$y_p(x) = A(x) + B(x) e^x$$

where A and B are functions of x given by the matrix equation

$$\begin{bmatrix} 1 & e^x \\ 0 & e^x \end{bmatrix} \begin{bmatrix} A'(x) \\ B'(x) \end{bmatrix} = \begin{bmatrix} 0 \\ \frac{1}{e^x+1} \end{bmatrix}$$

Solving this matrix equation (by multiplying both sides by the inverse of the matrix $\begin{bmatrix} 1 & e^x \\ 0 & e^x \end{bmatrix}$), we obtain

$$\begin{aligned} \begin{bmatrix} A'(x) \\ B'(x) \end{bmatrix} &= \begin{bmatrix} 1 & -1 \\ 0 & e^{-x} \end{bmatrix} \begin{bmatrix} 0 \\ \frac{1}{e^x+1} \end{bmatrix} \\ &= \begin{bmatrix} -\frac{1}{e^x+1} \\ \frac{e^{-x}}{e^x+1} \end{bmatrix} \end{aligned}$$

Hence,

$$\begin{aligned} A(x) &= -\int \frac{1}{e^x+1}dx \\ B(x) &= \int \frac{e^{-x}}{e^x+1}dx \end{aligned}$$

To determine $A(x)$, divide the top and bottom of the integrand by e^x :

$$\begin{aligned} A(x) &= -\int \frac{1}{e^x+1}dx \\ &= -\int \frac{e^{-x}}{1+e^{-x}}dx \end{aligned}$$

Let $u = 1+e^{-x}$, $du = -e^{-x}dx$. Hence,

$$\begin{aligned} A(x) &= -\int \frac{e^{-x}}{1+e^{-x}}dx \\ &= \int \frac{du}{u} \\ &= \ln|u| \\ &= \ln\left|1+e^{-x}\right| \\ &= \ln\left(1+e^{-x}\right) \end{aligned}$$

Similarly,

$$\begin{aligned} B(x) &= \int \frac{e^{-x}}{e^x+1}dx \\ &= \int \frac{e^{-2x}}{1+e^{-x}}dx \\ &= \int (\frac{1}{u}-1)du \qquad \left(u = 1+e^{-x}\right) \\ &= \ln|u| - u \\ &= \ln\left(1+e^{-x}\right) - \left(1+e^{-x}\right) \end{aligned}$$

Finally,

$$\begin{aligned} y_p(x) &= A(x) + B(x)e^x \\ &= \ln\left(1+e^{-x}\right) + e^x\left(\ln\left(1+e^{-x}\right) - \left(1+e^{-x}\right)\right) \\ &= (1+e^x)\ln\left(1+e^{-x}\right) - e^x\left(1+e^{-x}\right) \end{aligned}$$

The general solution of the differential equation is therefore given by

$$\begin{aligned} y(x) &= y_c(x) + y_p(x) \\ &= c_1 + c_2e^x + (1+e^x)\ln\left(1+e^{-x}\right) - e^x\left(1+e^{-x}\right) \\ &= c_3 + c_4e^x + (1+e^x)\ln\left(1+e^{-x}\right) \end{aligned}$$

where, $c_3 = c_1 - 1$ and $c_4 = c_2 - 1$.

4.(a)(10%)Classify all singular points *and* the point at infinity of the differential equation

$$\left(1-t^2\right)\frac{d^2y}{dt^2} - \frac{2t}{(1+t)^2}\frac{dy}{dt} + n(n+1)\,y = 0$$

where n is a positive integer.

(b)(20%) Find the general solution of the following differential equation in terms of power series about $x = 1$. State the region of validity of your solution (you must justify your conclusion).

$$y'' - (x-1)\,y' + 5y = 0$$

Here, y is a function of x.

Solution

Question 4.(a) Write the differential equation in the form

$$\frac{d^2y}{dt^2} - \left[\frac{2t}{(1-t^2)(1+t)^2}\right]\frac{dy}{dt} + \frac{n(n+1)}{(1-t^2)}y = 0$$

As in Section 3.8, we identify

$$\begin{aligned} P(t) &= -\frac{2t}{(1-t^2)(1+t)^2} \\ &= -\frac{2t}{(1-t)(1+t)^3} \\ Q(t) &= \frac{n(n+1)}{(1-t^2)} \\ &= \frac{n(n+1)}{(1-t)(1+t)} \end{aligned}$$

Since both $P(t)$ and $Q(t)$ fail to be analytic at $t = \pm 1$, we have singular points at $t = \pm 1$. We classify each of these singular points as follows (see Section 3.8).

$t = 1:$

$$\begin{aligned} p(t) &= (t-1)\,P(t) \\ &= -\frac{2t(t-1)}{(1-t)(1+t)^3} \\ &= \frac{2t}{(1+t)^3} \end{aligned}$$

$$\begin{aligned} q(t) &= (t-1)^2 Q(t) \\ &= (t-1)^2 \frac{n(n+1)}{(1-t)(1+t)} \\ &= \frac{n(n+1)(1-t)}{(1+t)} \end{aligned}$$

Since p and q are *both* analytic at $t = 1$, the latter is a *regular singular point.*

$t = -1$:

$$\begin{aligned} p(t) &= (t+1) P(t) \\ &= -\frac{2t(t+1)}{(1-t)(1+t)^3} \\ &= -\frac{2t}{(1-t)(1+t)^2} \end{aligned}$$

$$\begin{aligned} q(t) &= (t+1)^2 Q(t) \\ &= (t+1)^2 \frac{n(n+1)}{(1-t)(1+t)} \\ &= \frac{n(n+1)(1+t)}{(1-t)} \end{aligned}$$

Since p is *not* analytic at $t = -1$, the latter is an *irregular singular point.*

To identify the *point at infinity*, we make the substitution $w = \dfrac{1}{t}$ and classify the point $w = 0$ (see Note 3.8.9). The derivatives transform as follows.

$$\begin{aligned} \frac{dy}{dt} &= \frac{dy}{dw}\frac{dw}{dt} \\ &= \frac{dy}{dw}\left(-\frac{1}{t^2}\right) \\ &= -w^2\frac{dy}{dw} \\ \frac{d^2y}{dt^2} &= \frac{d}{dt}\left(\frac{dy}{dt}\right) \\ &= \frac{d}{dw}\left(\frac{dy}{dt}\right)\frac{dw}{dt} \\ &= \left(-w^2\frac{d^2y}{dw^2} - 2w\frac{dy}{dw}\right)\left(-w^2\right) \end{aligned}$$

The differential equation

$$\frac{d^2y}{dt^2} - \left[\frac{2t}{(1-t^2)(1+t)^2}\right]\frac{dy}{dt} + \frac{n(n+1)}{(1-t^2)}y = 0$$

becomes

$$\begin{aligned}
\left(-w^2\frac{d^2y}{dw^2} - 2w\frac{dy}{dw}\right)(-w^2) - \frac{\frac{2}{w}}{\left(1-\left(\frac{1}{w}\right)^2\right)\left(1+\frac{1}{w}\right)^2}\left(-w^2\frac{dy}{dw}\right) + \frac{n(n+1)}{\left(1-\left(\frac{1}{w}\right)^2\right)}y &= 0 \\
w^4\frac{d^2y}{dw^2} + 2\left[\frac{w^5}{(w^2-1)(w+1)^2} + w^3\right]\frac{dy}{dw} + \frac{n(n+1)w^2}{w^2-1}y &= 0 \\
\frac{d^2y}{dw^2} + 2\left[\frac{w}{(w^2-1)(w+1)^2} + \frac{1}{w}\right]\frac{dy}{dw} + \frac{n(n+1)}{w^2(w^2-1)}y &= 0
\end{aligned}$$

Consider the point $w = 0$ (or '$t = \infty$').

$$\begin{aligned}
P(w) &= 2\left[\frac{w}{(w^2-1)(w+1)^2} + \frac{1}{w}\right] \\
Q(w) &= \frac{n(n+1)}{w^2(w^2-1)}
\end{aligned}$$

Since both $P(w)$ and $Q(w)$ fail to be analytic at $w = 0$, the latter is a singular point. We classify this point as follows (see Section 3.8).

$$\begin{aligned}
p(w) &= wP(w) \\
&= 2w\left[\frac{w}{(w^2-1)(w+1)^2} + \frac{1}{w}\right] \\
&= 2\left[\frac{w^2}{(w^2-1)(w+1)^2} + 1\right]
\end{aligned}$$

$$\begin{aligned}
q(w) &= w^2Q(w) \\
&= w^2\frac{n(n+1)}{w^2(w^2-1)} \\
&= \frac{n(n+1)}{(w^2-1)}
\end{aligned}$$

Since p and q are *both* analytic at $w = 0$, the latter and hence $t = \infty$ (the point at infinity of our original equation) is a *regular singular point.*

Question 4.(b) In accordance with Note 3.8.5, let $u = x - 1$ and examine the resulting differential equation near $u = 0$.

$$\begin{aligned}
y'' - (x-1)y' + 5y &= 0 \\
\frac{d^2y}{du^2} - u\frac{dy}{du} + 5y &= 0 \qquad \text{(F4.6)}
\end{aligned}$$

Since $P(u) = -u$ and $Q(u) = 5$ are both analytic at $u = 0$, the latter is an *ordinary* point. In fact, the equation (F4.6) has no singular points. Consequently, by Theorem 3.8.3, we can represent the general solution of (F4.6) in the form

$$y(u) = \sum_{n=0}^{\infty} b_n u^n \,, \quad b_0,\ b_1 \ \text{ arbitrary constants} \tag{F4.7}$$

This solution will be valid for *all* u. To find the b_n's, we substitute (F4.7) into (F4.6). To this end, let

$$\begin{aligned} y(u) &= \sum_{n=0}^{\infty} b_n u^n \\ y'(u) &= \sum_{n=0}^{\infty} n b_n u^{n-1} \\ y''(u) &= \sum_{n=0}^{\infty} n(n-1) b_n u^{n-2} \end{aligned}$$

Substitute these relations into (F4.6).

$$\begin{aligned} \sum_{n=0}^{\infty} n(n-1) b_n u^{n-2} - u \sum_{n=0}^{\infty} n b_n u^{n-1} + 5 \sum_{n=0}^{\infty} b_n u^n &= 0 \\ \sum_{n=2}^{\infty} n(n-1) b_n u^{n-2} - \sum_{n=0}^{\infty} n b_n u^n + 5 \sum_{n=0}^{\infty} b_n u^n &= 0 \end{aligned}$$

Shifting the summation in the first sum on the left-hand side ($n \longrightarrow n+2$), we obtain

$$\sum_{n=0}^{\infty} [(n+2)(n+1) b_{n+2} - n b_n + 5\, b_n] u^n = 0$$

Comparing coefficients of u^n on both sides of this equation, we obtain

$$\begin{aligned} (n+2)(n+1) b_{n+2} + b_n (5-n) &= 0 \\ b_{n+2} &= \frac{(n-5)\, b_n}{(n+1)(n+2)} \,, \quad n \geq 0 \end{aligned}$$

The equation

$$b_{n+2} = \frac{(n-5)\, b_n}{(n+1)(n+2)} \,, \quad n \geq 0 \tag{F4.8}$$

is a two-term recurrence relation for the coefficients b_n , $n \geq 0$ (leading to two arbitrary constants - as expected from Theorem 3.8.3.). Let b_0 and b_1 be arbitrary constants. From (F4.8),

$$b_0 = b_0 \qquad b_1 = b_1$$

$$b_2 = \tfrac{-5b_0}{1\cdot 2} \qquad b_3 = \tfrac{-4b_1}{2\cdot 3}$$

$$b_4 = \tfrac{-3b_2}{3\cdot 4} = \tfrac{-3(-5)b_0}{4!} \qquad b_5 = \tfrac{-2b_3}{4\cdot 5} = \tfrac{(-2)(-4)b_1}{5!}$$

$$b_6 = \tfrac{-1b_4}{5\cdot 6} = \tfrac{-1(-3)(-5)b_0}{6!} \qquad b_7 = 0$$

$$\vdots \qquad \vdots$$

$$b_{2n} = \tfrac{(2n-7)\cdot\ldots\cdot(-3)(-5)}{(2n)!}b_0\,, \quad n \geq 1 \qquad b_{2n+1} = 0\,, \quad n \geq 3$$

Hence, the general solution of (F4.6) is given by

$$\begin{aligned} y(u) &= \sum_{n=0}^{\infty} b_n u^n \\ &= \sum_{n=0}^{\infty} b_{2n} u^{2n} + \sum_{n=0}^{\infty} b_{2n+1} u^{2n+1} \\ &= b_0 + \sum_{n=1}^{\infty} \frac{(2n-7)\cdot \ldots\ldots \cdot (-3)\,(-5)}{(2n)!} b_0 u^{2n} + b_1 u - \frac{2b_1}{3}u^3 + \frac{b_1}{15}u^5 \\ &= b_0 \sum_{n=0}^{\infty} \frac{-15u^{2n}}{2^n n!\,(2n-1)\,(2n-3)\,(2n-5)} + b_1\left(u - \frac{2u^3}{3} + \frac{u^5}{15}\right) \end{aligned}$$

valid for all u (certainly 'near $u = 0$' as required). Finally, let $u = x - 1$, to obtain the general solution

$$y(x) = b_0 \sum_{n=0}^{\infty} \frac{-15\,(x-1)^{2n}}{2^n n!\,(2n-1)\,(2n-3)\,(2n-5)} + b_1\left[(x-1) - \frac{2\,(x-1)^3}{3} + \frac{(x-1)^5}{15}\right]$$

valid for all x (certainly 'near $x = 1$', as required).

5. Find the following
(a)(5%) $L\left[t^2 \cos t\right]$
(b)(7%) $L[F(t)]$ where

$$F(t) = \begin{cases} 0, & 0 \leq t \leq 1, \\ \cos \pi t, & 2 < t < 3, \\ 0, & t \geq 3 \end{cases}$$

Solution

Question 5.(a) From Table 3.9.2,

$$L\left[t^n y(t)\right] = (-1)^n \frac{d^n f}{ds^n}(s)$$

where

$$f(s) = L[y(t)]$$

Hence,

$$L\left[t^2 \cos t\right] = (-1)^2 \frac{d^2 f}{ds^2}$$

where

$$f(s) = L[\cos t] = \frac{s}{s^2+1}$$

Differentiating $f(s)$ with respect to s, we obtain

$$\begin{aligned} \frac{df}{ds} &= \frac{d}{ds}\left(\frac{s}{s^2+1}\right) \\ &= \frac{1-s^2}{\left(s^2+1\right)^2} \\ \frac{d^2 f}{ds^2} &= \frac{-2s\left(s^2+1\right)^2 - 2\left(s^2+1\right)(2s)\left(1-s^2\right)}{\left(s^2+1\right)^4} \\ &= \frac{2s\left(s^2-3\right)}{\left(s^2+1\right)^3} \end{aligned}$$

Finally,

$$\begin{aligned} L\left[t^2 \cos t\right] &= (-1)^2 \frac{d^2 f}{ds^2} \\ &= \frac{2s\left(s^2-3\right)}{\left(s^2+1\right)^3} \end{aligned}$$

Question 5.(b) First write $F(t)$ in terms of Heaviside functions

$$\begin{aligned} F(t) &= \begin{cases} 0, & 0 \le t < 2, \\ \cos \pi t, & 2 \le t < 3, \\ 0, & t \ge 3 \end{cases} \\ &= \begin{cases} 0, & 0 \le t < 2 \\ \cos \pi t, & t \ge 2 \end{cases} - \begin{cases} 0, & 0 \le t < 3 \\ \cos \pi t, & t \ge 3 \end{cases} \\ &= \left[\alpha(t-2) - \alpha(t-3)\right] \cos \pi t \end{aligned}$$

From Table 3.9.2,

$$L[y(t-c)\,\alpha(t-c)] = e^{-cs} f(s) \tag{F4.9}$$

where $f(s) = L[y(t)]$. To get the required form, we note that, since $\cos(x+y) = \cos x \cos y - \sin x \sin y$,

$$\begin{aligned}\alpha(t-2)\cos \pi t &= \alpha(t-2)\cos \pi(t-2) \\ \alpha(t-3)\cos \pi t &= -\alpha(t-3)\cos \pi(t-3)\end{aligned}$$

Hence, from (F4.9),

$$\begin{aligned}L[F(t)] &= L[\{\alpha(t-2) - \alpha(t-3)\}\cos \pi t] \\ &= L[\alpha(t-2)\cos \pi(t-2)] - L[\alpha(t-3)\cos \pi(t-3)] \\ &= e^{-2s}L[\cos \pi t] - e^{-3s}L[\cos \pi t] \\ &= \frac{s}{s^2+1}\left(e^{-2s} - e^{-3s}\right)\end{aligned}$$

Question 6.(8%) Use the derivative of the function

$$f(s) = \ln\left(2 + \frac{3}{s}\right), \quad s > 0$$

to find its inverse Laplace transform.

Solution. Let $f(s) = \ln\left(2 + \frac{3}{s}\right)$ so that $y(t) = L^{-1}[f(s)]$. Hence,

$$\begin{aligned}\frac{df}{ds} &= \frac{1}{2+\frac{3}{s}}\left(-\frac{3}{s^2}\right) \\ &= -\frac{3}{s(2s+3)}\end{aligned}$$

From Table 3.9.2

$$t^n y(t) = L^{-1}[(-1)^n \frac{d^n f}{ds^n}(s)]$$

Thus,

$$\begin{aligned}y(t) &= L^{-1}[f(s)] \\ &= \frac{1}{t}L^{-1}\left[(-1)\frac{df}{ds}\right] \\ &= -\frac{1}{t}L^{-1}\left[-\frac{3}{s(2s+3)}\right] \\ &= \frac{1}{t}L^{-1}\left[\frac{3}{s(2s+3)}\right]\end{aligned}$$

Now use *partial fractions* to invert the transform.

$$
\begin{aligned}
y(t) &= \frac{1}{t} L^{-1} [\frac{1}{s} - \frac{2}{2s+3}] \\
&= \frac{1}{t} (1 - L^{-1} \left[\frac{1}{s + \frac{3}{2}} \right]) \\
&= \frac{1}{t} \left(1 - e^{-\frac{3}{2}t} \right)
\end{aligned}
$$

Question 7.(15%) Find a series solution of the following differential equation near $x = 0$. What form should the other linearly independent solution take (you must justify your conclusion) ?

$$
x \frac{d^2y}{dx^2} + (1 - x) \frac{dy}{dx} + \alpha y = 0, \quad x > 0, \quad \alpha \text{ is a positive integer}
$$

Solution. From Section 3.8, $x = 0$ is an *ordinary* point of the differential equation if the expressions $P(x)$ and $Q(x)$ in (3.8.4) of Chapter 3 are *both* analytic at $x = 0$. In fact, here

$$
\begin{aligned}
P(x) &= \frac{1 - x}{x} \\
Q(x) &= \frac{\alpha}{x}
\end{aligned}
$$

It is clear that neither $P(x)$ nor $Q(x)$ are analytic at $x = 0$ so that $x = 0$ is a *singular* point of the differential equation. According to the theory in Section 3.8, $x = 0$ is a *regular singular point* if $p(x) = xP(x)$ and $q(x) = x^2 Q(x)$ are *both* analytic at $x = 0$. In fact, here,

$$
\begin{aligned}
p(x) &= xP(x) = 1 - x \\
q(x) &= x^2 Q(x) = x\alpha
\end{aligned}
$$

which are indeed *both* analytic at $x = 0$. Hence, $x = 0$ is a *regular singular point* and, by Theorem 3.8.6, for $x > 0$, the differential equation has at least one Frobenius series solution (corresponding to the larger root of the indicial equation (3.8.13)). To see which form a *second linearly independent solution* will take, we examine the roots of the indicial equation (3.8.13) .

$$
\begin{aligned}
r(r-1) + p(0)r + q(0) &= 0 \\
r(r-1) + r &= 0 \\
r^2 &= 0 \\
r &= 0 \quad \text{(twice)}
\end{aligned}
$$

The roots of the indicial equation are such that $r_1 = r_2 = 0$. Hence, by Theorem 3.8.6(ii), the first solution is a Frobenius (in fact, power) series

$$y_1(x) = \sum_{n=0}^{\infty} a_n x^n \ , \qquad a_0 \neq 0$$

while the second linearly independent solution takes the form

$$y_2(x) = y_1(x) \ln x + \sum_{n=0}^{\infty} b_n x^{n+1}$$

To find the series solution $y_1(x)$,we substitute the series

$$y_1(x) = \sum_{n=0}^{\infty} a_n x^n \ , \qquad a_0 \neq 0$$

into the differential equation

$$x\frac{d^2y}{dx^2} + (1-x)\frac{dy}{dx} + \alpha y = 0$$

This gives

$$\begin{aligned}
\sum_{n=0}^{\infty} a_n n(n-1)x^{n-1} + (1-x)\sum_{n=0}^{\infty} a_n n x^{n-1} + \alpha \sum_{n=0}^{\infty} a_n x^n &= 0 \\
\sum_{n=0}^{\infty} a_n n(n-1)x^{n-1} + \sum_{n=0}^{\infty} a_n n x^{n-1} - \sum_{n=0}^{\infty} a_n n x^n + \alpha \sum_{n=0}^{\infty} a_n x^n &= 0
\end{aligned}$$

Simplifying and shifting summation index in the third and fourth sums on the left-hand side ($n \to n-1$), we obtain

$$\begin{aligned}
\sum_{n=0}^{\infty} (a_n n(n-1) + a_n n)x^{n-1} - \sum_{n=1}^{\infty} (a_{n-1}(n-1) + \alpha a_{n-1})x^{n-1} &= 0 \\
\sum_{n=0}^{\infty} n^2 a_n x^{n-1} - \sum_{n=1}^{\infty} a_{n-1}(n-1+\alpha)x^{n-1} &= 0
\end{aligned}$$

Equate coefficients of x^{n-1} for different values of n :

$$n = 0: \quad a_0 \cdot 0 = 0, \qquad a_0 \neq 0 \quad \therefore \ a_0 = \text{arbitrary}$$

$$n \geq 1: \quad a_n = \frac{n-1+\alpha}{n^2} a_{n-1}$$

From this recurrence relation for $n \geq 1$, we obtain

$$\begin{aligned}
a_0 &= a_0 \\
a_1 &= \alpha a_0 \\
a_2 &= \frac{(1+\alpha)a_1}{2^2} = \frac{\alpha(1+\alpha)a_0}{2^2} \\
a_3 &= \frac{(2+\alpha)a_2}{3^2} = \frac{\alpha(1+\alpha)(2+\alpha)a_0}{2^2 \cdot 3^2} \\
&\vdots \\
&\vdots \\
a_n &= \frac{\alpha(1+\alpha)\cdot \ldots \cdot(n-1+\alpha)}{2^2 \cdot 3^2 \cdot \ldots \cdot n^2} a_0 \,, \quad n \geq 1
\end{aligned}$$

Hence, the series solution is given by

$$\begin{aligned}
y(x) &= \sum_{n=0}^{\infty} a_n x^n \\
&= a_0 [1 + \sum_{n=1}^{\infty} \frac{\alpha\,(1+\alpha)\cdot \ldots \cdot(n-1+\alpha)}{2^2 \cdot 3^2 \cdot \ldots \cdot n^2} x^n]
\end{aligned}$$

Since there are no other singular points, this solution is valid for all $x > 0$.

FINAL EXAMINATION #5
SOLUTIONS

In what follows, c_i, $i = 1, 2, ...$ will denote arbitrary constants. Rules of logarithms and exponentials, a summary of the main techniques of integration as well as a *table of integrals* can be found in the Appendix.

Question 1.(25%) Find the general solution, valid near $x = 0$, of Bessel's equation of order 1

$$x^2\frac{d^2y}{dx^2} + x\frac{dy}{dx} + (x^2 - 1)\,y = 0, \qquad x > 0$$

Solution. From Section 3.8, $x = 0$ is an *ordinary* point of the differential equation if the expressions $P(x)$ and $Q(x)$ in (3.8.4) of Chapter 3 are *both* analytic at $x = 0$. In fact, here

$$\begin{aligned} P(x) &= \frac{x}{x^2} \\ Q(x) &= \frac{x^2 - 1}{x^2} \end{aligned}$$

It is clear that neither $P(x)$ nor $Q(x)$ are analytic at $x = 0$ so that $x = 0$ is a *singular* point of the differential equation. According to the theory in Section 3.8, $x = 0$ is a *regular singular point* if $p(x) = xP(x)$ and $q(x) = x^2Q(x)$ are *both* analytic at $x = 0$. In fact, here,

$$\begin{aligned} p(x) &= xP(x) = 1 \\ q(x) &= x^2Q(x) = x^2 - 1 \end{aligned}$$

which are indeed *both* analytic at $x = 0$. Hence, $x = 0$ is a *regular singular point* and, by Theorem 3.8.6, for $x > 0$, the differential equation has at least one Frobenius series solution (corresponding to the larger root of the indicial equation (3.8.13)). To see which form a *second linearly independent solution* will take, we examine the roots of the indicial equation (3.8.13) .

$$\begin{aligned} r(r-1) + p(0)r + q(0) &= 0 \\ r(r-1) + r - 1 &= 0 \\ r^2 - 1 &= 0 \\ (r-1)(r+1) &= 0 \\ r_1 &= 1, \quad r_2 = -1 \end{aligned}$$

The roots of the indicial equation are such that $r_1 - r_2$ is a positive integer. Hence, by Theorem 3.8.6(iii), a second linearly independent solution of the differential equation takes the form

$$y_2(x) = Cy_1(x)\ln x + \sum_{n=0}^{\infty} b_n x^{n+r_2} \tag{F5.1}$$

where $b_0 \neq 0$ but the constant C may be either zero or nonzero and $y_1(x)$ is the Frobenius series solution corresponding to the larger root $r_1 = 1$:

$$y_1(x) = \sum_{n=0}^{\infty} a_n x^{n+1} \quad , \qquad a_0 \neq 0$$

Hence, the second linearly independent solution may or may not be a Frobenius series (depending on whether the logarithmic term in (F5.1) vanishes, that is, whether or not $C = 0$). To see which is the case, we proceed as in Final Exam #3, Q4(i) and use the smaller root $r_2 = -1$ of the indicial equation to obtain a recurrence relation. If a second linearly independent Frobenius solution exists, this procedure will yield *both* Frobenius series solutions (those corresponding to r_2 *and* r_1) *at the same time.* If it doesn't, we know that the second linearly independent solution is not a Frobenius series and does, in fact, involve a logarithmic term. Hence, substitute the Frobenius series

$$y(x) = \sum_{n=0}^{\infty} b_n x^{n-1} \quad , \qquad b_0 \neq 0$$

into the differential equation

$$x^2 \frac{d^2y}{dx^2} + x\frac{dy}{dx} + \left(x^2 - 1\right) y = 0$$

This gives

$$\sum_{n=0}^{\infty} b_n (n-1)(n-1)x^{n-1} + \sum_{n=0}^{\infty} b_n (n-1)\, x^{n-1} + \sum_{n=0}^{\infty} b_n x^{n+1} - \sum_{n=0}^{\infty} b_n x^{n-1} = 0$$

Simplifying and shifting summation index in the third sum ($n \to n-2$), we obtain

$$\sum_{n=0}^{\infty} b_n n\,(n-2)\, x^{n-1} + \sum_{n=2}^{\infty} b_{n-2} x^{n-1} = 0$$

Equate coefficients of x^{n-1} for different values of n :

$$\begin{aligned} n &= 0: \; b_0 \cdot (0) \cdot (-2) = 0, \qquad b_0 \neq 0 \Rightarrow b_0 \text{ is arbitrary} \\ n &= 1: \; b_1 \cdot (-1) = 0 \Rightarrow b_1 = 0 \\ n &\geq 2: \; b_n = -\frac{b_{n-2}}{n\,(n-2)} \end{aligned}$$

Hence, we have only one arbitrary constant b_0 so that only *one* Frobenius series solution exists (contrast with Final Exam #3, Q4(i)). This means that $C \neq 0$ in (F5.1) so that the second linearly independent solution is, in fact, logarithmic. To obtain both linearly independent solutions ((F5.1) *and* the Frobenius series solution

$$y_1(x) = \sum_{n=0}^{\infty} a_n x^{n+1}, \qquad a_0 \neq 0$$

corresponding to the larger root $r = r_1 = 1$), we proceed as follows.

First substitute the general series

$$y(x, r) = \sum_{n=0}^{\infty} c_n x^{n+r}$$

into the differential equation. This gives

$$\sum_{n=0}^{\infty} c_n (n+r)(n+r-1) x^{n+r} + \sum_{n=0}^{\infty} c_n (n+r) x^{n+r} + \sum_{n=0}^{\infty} c_n x^{n+r+2} - \sum_{n=0}^{\infty} c_n x^{n+r} = 0$$

Simplifying and shifting summation index in the third sum ($n \to n-2$), we obtain

$$\sum_{n=0}^{\infty} c_n[(n+r)^2 - 1]x^{n+r} + \sum_{n=2}^{\infty} c_{n-2} x^{n+r} = 0$$

Equate coefficients of x^{n+r} for different values of n :

$$\begin{aligned}
n &= 0: \ c_0 \cdot \left(r^2 - 1\right) = 0, \qquad c_0 \neq 0 \Rightarrow r^2 - 1 = 0: \ \text{indicial equation} \\
n &= 1: \ c_1 \cdot \left[(1+r)^2 - 1\right] = 0 \Rightarrow c_1 = 0 \text{ for } r = \pm 1 \\
n &\geq 2: \ \left[(n+r)^2 - 1\right] c_n = -c_{n-2}
\end{aligned}$$

Consider the recurrence relation for $r \neq r_2 = -1$. We have

$$\begin{aligned}
c_n &= -\frac{c_{n-2}}{(n+r)^2 - 1} \\
&= -\frac{c_{n-2}}{(n+r+1)(n+r-1)}, \quad n \geq 2
\end{aligned}$$

Thus,

$$\begin{aligned}
c_0 &= c_0 & c_1 &= 0 \\
c_2 &= -\frac{c_0}{(3+r)(1+r)} & c_3 &= 0 \\
c_4 &= -\frac{c_2}{(r+3)(r+5)} = \frac{c_0}{(r+1)(r+3)^2(r+5)} & c_5 &\ 0 \\
c_6 &= -\frac{c_4}{(r+7)(r+5)} = -\frac{c_0}{[(r+1)(r+3)(r+5)][(r+3)(r+5)(r+7)]} & c_7 &= 0 \\
&\vdots & &\vdots \\
c_{2n} &= \frac{(-1)^n}{[(r+1)(r+3)\ldots(r+2n-1)][(r+3)(r+5)\ldots(r+2n+1)]} c_0, \quad n \geq 1 & c_{2n+1} &= 0, \quad n \geq 0
\end{aligned}$$

We let $c_0 = r - r_2 = r + 1$ so that

$$c_{2n} = \frac{(-1)^n (r+1)}{[(r+1)(r+3)\ldots(r+2n-1)][(r+3)(r+5)\ldots(r+2n+1)]}, \quad n \geq 1$$

and

$$\begin{aligned} y(x,r) &= \sum_{n=0}^{\infty} c_n x^{n+r} \\ &= (r+1)x^r + \sum_{n=1}^{\infty} \frac{(-1)^n (r+1) x^{2n+r}}{[(r+1)\ldots(r+2n-1)][(r+3)\ldots(r+2n+1)]} \\ &= (r+1)x^r - \frac{(r+1)x^{r+2}}{(r+1)(r+3)} + \sum_{n=2}^{\infty} \frac{x^{2n+r}(-1)^n}{[(r+3)\ldots(r+2n-1)][(r+3)\ldots(r+2n+} \\ &= (r+1)x^r - \frac{x^{r+2}}{(r+3)} + \sum_{n=2}^{\infty} \frac{x^{2n+r}(-1)^n}{[(r+3)\ldots(r+2n-1)][(r+3)\ldots(r+2n+1)]} \quad \text{(F} \end{aligned}$$

The first solution is now obtained from

$$\begin{aligned} y_1(x) &= [y(x,r)]_{r=r_2=-1} \\ &= 0 - \frac{1}{2}x + \sum_{n=2}^{\infty} \frac{(-1)^n x^{2n-1}}{[2\cdot 4\cdot \ldots \cdot 2n-2][2\cdot 4\cdot \ldots \cdot 2n]} \\ &= -\frac{1}{2}x + \sum_{n=2}^{\infty} \frac{(-1)^n x^{2n-1}}{2^{n-1}(n-1)!2^n n!} \end{aligned}$$

To recognize this series as the Frobenius series corresponding to the larger root $r = r_1 = 1$, we let $n \to n+1$ and obtain

$$\begin{aligned} y_1(x) &= -\frac{1}{2}x + \sum_{n=1}^{\infty} \frac{(-1)^{n+1} x^{2n+1}}{2^{2n+1} n!(n+1)!} \\ &= -\frac{x}{2} \sum_{n=0}^{\infty} \frac{(-1)^n x^{2n}}{2^{2n} n!(n+1)!}, \quad x > 0 \end{aligned}$$

The second linearly independent solution is also obtained from the smaller root $r = r_2 = -1$ of the indicial equation as follows. From (F5.2),

$$\begin{aligned}
\frac{\partial}{\partial r}y(x,r) &= \frac{\partial}{\partial r}\{(r+1)x^r - \frac{x^{r+2}}{r+3} + \sum_{n=2}^{\infty}\frac{(-1)^n\ x^{2n+r}}{[(r+3)\ldots(r+2n-1)][(r+3)\ldots(r+2n+1)]}\} \\
&= x^r + (r+1)x^r\ln x - \frac{x^{r+2}\ln x}{r+3} + \frac{x^{r+2}}{(r+3)^2} \\
&\quad + \sum_{n=2}^{\infty}\frac{(-1)^n x^{2n+r}\ln x}{[(r+3)\ldots(r+2n-1)][(r+3)\ldots(r+2n+1)]} \\
&\quad + \sum_{n=2}^{\infty}\{\frac{(-1)^n x^{2n+r}\ln x}{[(r+3)\ldots(r+2n-1)][(r+3)\ldots(r+2n+1)]} \\
&\quad \times(-1)\left[\frac{1}{r+3}+\ldots+\frac{1}{r+2n+1}+\frac{1}{r+3}+\ldots+\frac{1}{r+2n-1}\right]\}
\end{aligned}$$

(Here, we have used the formula for the differentiation of a product of functions i.e. if

$$f(r) = f_1(r)\,f_2(r)\ldots f_n(r)$$

then

$$f'(r) = f\{\frac{f_1'}{f_1}+\frac{f_2'}{f_2}+\ldots+\frac{f_n'}{f_n}\}$$

where " $'$ " denotes differentiation with respect to r.). The second linearly independent solution is now given by

$$\begin{aligned}
y_2(x) &= \left[\frac{\partial}{\partial r}y(x,r)\right]_{r=r_2=-1} \\
&= x^{-1} - \frac{x\ln x}{2} + \frac{x}{4} + \ln x\sum_{n=2}^{\infty}\frac{(-1)^n x^{2n-1}}{[2\cdot4\cdot\ldots\cdot2n-2]\,[2\cdot4\cdot\ldots\cdot2n]} \\
&\quad + \sum_{n=2}^{\infty}\frac{(-1)^n x^{2n-1}}{[2\cdot4\cdot\ldots\cdot2n-2]\,[2\cdot4\cdot\ldots\cdot2n]} \\
&\quad \times\{-\frac{1}{2}-\ldots-\frac{1}{2n}-\frac{1}{2}-\ldots-\frac{1}{2n-2}\} \\
&= \left[-\frac{x}{2}+\sum_{n=2}^{\infty}\frac{(-1)^n x^{2n-1}}{[2\cdot4\cdot\ldots\cdot2n-2]\,[2\cdot4\cdot\ldots\cdot2n]}\right]\ln x \\
&\quad + x^{-1} + \frac{x}{4} - \sum_{n=2}^{\infty}\frac{(-1)^n\ x^{2n-1}}{[2\cdot4\cdot\ldots\cdot2n-2]\,[2\cdot4\cdot\ldots\cdot2n]}\left(\frac{1}{2}\sum_{i=1}^{n}\frac{1}{i}+\frac{1}{2}\sum_{i=1}^{n-1}\frac{1}{i}\right)
\end{aligned}$$

Simplifying this expression and noting the form of $y_1(x)$ above, we obtain

$$\begin{aligned} y_2(x) &= y_1 \ln x + x^{-1} + \frac{x}{4} - \sum_{n=2}^{\infty} \frac{(-1)^n x^{2n-1}}{2^{2n} n! (n-1)!} (H_n + H_{n-1}) \\ &= y_1 \ln x + x^{-1} - \sum_{n=1}^{\infty} \frac{(-1)^n x^{2n-1}}{2^{2n} n! (n-1)!} (H_n + H_{n-1}), \quad x > 0 \end{aligned}$$

Here,

$$H_n = \sum_{i=1}^{n} \frac{1}{i}$$

denotes the nth partial sum of the harmonic series (Notice how both linearly independent solutions are again obtained from the smaller root $r = r_2 = -1$).

Finally, the general solution is given by

$$y(x) = c_1 y_1(x) + c_2 y_2(x)$$

valid in the region $x > 0$ (since the differential equation has no other singular points).

Question 2.(15%) Use the Laplace transform method to solve the following initial value problem.

$$\begin{aligned} \frac{d^2x}{dt^2} + 3\frac{dx}{dt} &= H(t) \\ x(0) &= \frac{dx}{dt}(0) = 0 \end{aligned}$$

where

$$H(t) = \begin{cases} \cos 3t, & 0 \leq t < 3\pi, \\ 1 + \cos 3t, & t \geq 3\pi \end{cases}$$

Solution. First take the Laplace transform of both sides of the differential equation. From Table 3.9.2,

$$\begin{aligned} L[\frac{d^2x}{dt^2} + 3\frac{dx}{dt}] &= L[H(t)] \\ s^2 L[x] - sx(0) - \frac{dx}{dt}(0) + 3(sL[x] - x(0)) &= L[H(t)] \end{aligned}$$

Applying the given initial conditions, we obtain

$$(s^2 + 3s) L[x] = L[H(t)] \tag{F5.3}$$

To find the Laplace transform of the function $H(t)$, we first write it in terms of the Heaviside function $\alpha(t-c)$ (see Table 3.9.2).

$$\begin{aligned} H(t) &= \cos 3t + \begin{cases} 0, & 0 \le t < 3\pi, \\ 1, & t \ge 3\pi \end{cases} \\ &= \cos 3t + \alpha(t - 3\pi) \end{aligned}$$

Hence, from Table 3.9.1,

$$\begin{aligned} L[H(t)] &= L[\cos 3t] - \underbrace{L[\alpha(t-3\pi)]}_{\text{Table 3.9.2}} \\ &= \frac{s}{s^2+9} - \frac{e^{-3\pi s}}{s} \end{aligned}$$

From (F5.3),

$$\begin{aligned} \left(s^2+3\right) L[x] &= \frac{s}{s^2+9} - \frac{e^{-3\pi s}}{s} \\ L[x] &= \frac{s}{(s^2+9)(s^2+3)} - \frac{e^{-3\pi s}}{s(s^2+3)} \end{aligned}$$

Using partial fractions (see Appendix), we obtain

$$L[x] = -\frac{1}{6}\left[\frac{s}{s^2+9} - \frac{s}{s^2+3}\right] - \frac{e^{-3\pi s}}{3}\left(\frac{1}{s} - \frac{s}{s^2+3}\right)$$

Inverting the Laplace transform and using Tables 3.9.1 and 3.9.2, we have

$$\begin{aligned} x(t) &= -\frac{1}{6}L^{-1}\left[\frac{s}{s^2+9}\right] + \frac{1}{6}L^{-1}\left[\frac{s}{s^2+3}\right] - \frac{1}{3}L^{-1}\left[\frac{e^{-3\pi s}}{s}\right] + \frac{1}{3}L^{-1}\left[\frac{se^{-3\pi s}}{s^2+3}\right] \\ &= -\frac{1}{6}\cos 3t + \frac{1}{6}\sin\left(\sqrt{3}t\right) - \frac{1}{3}\alpha(t-3\pi) + \frac{1}{3}\alpha(t-3\pi)F_1(t-3\pi) \qquad \text{(F5.4)} \end{aligned}$$

where

$$\begin{aligned} F_1(t) &= L^{-1}\left[\frac{s}{s^2+3}\right] \\ &= \cos\sqrt{3}t \end{aligned}$$

Hence,

$$F_1(t-3\pi) = \cos\left[\sqrt{3}(t-3\pi)\right]$$

Finally, from (F5.4),

$$\begin{aligned} x(t) &= -\frac{1}{6}\cos 3t + \frac{1}{6}\sin\left(\sqrt{3}t\right) - \frac{1}{3}\alpha(t-3\pi) + \frac{1}{3}\alpha(t-3\pi)F_1(t-3\pi) \\ &= -\frac{1}{6}[\cos 3t - \sin\left(\sqrt{3}t\right)] - \frac{1}{3}\{\alpha(t-3\pi) - \alpha(t-3\pi)\cos\left[\sqrt{3}(t-3\pi)\right]\} \end{aligned}$$

Question 3.(15%) Use the *method of undetermined coefficients* to find the general solution of

$$\frac{d^2y}{dx^2} + 4y = 2\sin^2 x$$

Which other method is suitable in this case ? Justify your conclusion.

Solution. To use the *method of undetermined coefficients*, we must first write the rightp-hand side of the differential equation in a form suitable for Table 3.3.1. Since $\cos 2x = 1 - 2\sin^2 x$,

$$2\sin^2 x = 1 - \cos 2x$$

Hence, the differential equation becomes

$$\frac{d^2y}{dx^2} + 4y = 1 - \cos 2x$$

Since the equation is linear, its general solution $y(x)$ can be split into two parts (see Section 3.1):

$$y(x) = y_c(x) + y_p(x)$$

For $y_c(x)$, the characteristic equation is

$$\begin{aligned} m^2 + 4 &= 0 \\ m &= \pm 2i \end{aligned}$$

From Table 3.2.1,

$$y_c(x) = c_1 \cos 2x + c_2 \sin 2x$$

For $y_p(x)$, from Table 3.3.1, we suggest

$$\begin{aligned} y_p(x) &= y_{p_1}(x) + y_{p_2}(x) \\ &= A + x(B\cos 2x + C\sin 2x) \end{aligned}$$

where A, B and C are definite constants to be determined. Note the need to multiply our initial guess for $y_{p_2}(x)$ (that corresponding to the term $\cos 2x$ on the right-hand side of the differential equation) by a factor of x to avoid terms in common with y_c. To find A, B and C, we substitute $y_p(x)$ into the differential equation.

$$\begin{aligned} y_p'' + 4y_p &= 1 - \cos 2x \\ -4B\sin 2x + 4C\cos 2x + 4A &= 1 - \cos 2x \end{aligned}$$

We obtain the following system of linear equations for A and B.

$$\begin{aligned} -4B &= 0 \\ 4C &= -1 \\ 4A &= 1 \end{aligned}$$

Hence,

$$A=\frac{1}{4},\quad B=0,\quad C=-\frac{1}{4}$$

so that

$$y_p(x)=\frac{1}{4}(1-x\sin 2x)$$

Finally, the general solution of the differential equation is given by

$$\begin{aligned} y(x) &= y_c(x)+y_p(x) \\ &= c_1\cos 2x+c_2\sin 2x+\frac{1}{4}(1-x\sin 2x) \end{aligned}$$

This problem can also be solved using either *Reduction of Order* (Section 3.4) or *Variation of Parameters* (Section 3.5) since we know the general solution $y_c(x)$ of the homogeneous equation.

Question 4.(10%) Find $L[|\sin kt|]$ where k is a non-zero integer.

Solution. First we note that $|\sin kt|$ is periodic with period $\frac{\pi}{k}$. That is

$$\left|\sin\left(kt+\frac{\pi}{k}\right)\right|=|\sin kt|,\quad 0\le t\le\frac{\pi}{k}$$

Hence, from Table 3.9.2,

$$\begin{aligned} L[|\sin kt|] &= \frac{\int_0^{\frac{\pi}{k}} e^{-st}|\sin kt|\,dt}{1-e^{-\frac{s\pi}{k}}} \\ &= \frac{\int_0^{\frac{\pi}{k}} e^{-st}\sin ktdt}{1-e^{-\frac{s\pi}{k}}} \end{aligned}$$

To evaluate the integral, we use integration by parts (or a table of integrals - see Appendix) and obtain

$$\int_0^{\frac{\pi}{k}} e^{-st}\sin ktdt=\frac{ke^{-\frac{s\pi}{k}}+k}{s^2+k^2}$$

Hence,

$$\begin{aligned} L[|\sin kt|] &= \frac{\int_0^{\frac{\pi}{k}} e^{-st}\sin ktdt}{1-e^{-\frac{s\pi}{k}}} \\ &= \frac{1}{1-e^{-\frac{s\pi}{k}}}\left[\frac{ke^{-\frac{s\pi}{k}}+k}{s^2+k^2}\right] \\ &= \frac{k}{s^2+k^2}\left[\frac{1+e^{-\frac{s\pi}{k}}}{1-e^{-\frac{s\pi}{k}}}\right] \end{aligned}$$

Multiplying the numerator and the denominator of the expression $\frac{1+e^{-\frac{s\pi}{k}}}{1-e^{-\frac{s\pi}{k}}}$ by $e^{-\frac{s\pi}{2k}}$, and noting that

$$\begin{aligned} \cosh x &= \frac{e^x + e^{-x}}{2} \\ \sinh x &= \frac{e^x - e^{-x}}{2} \end{aligned}$$

we obtain

$$\begin{aligned} \frac{1+e^{-\frac{s\pi}{k}}}{1-e^{-\frac{s\pi}{k}}} &= \frac{e^{-\frac{s\pi}{2k}} + e^{-\frac{s\pi}{2k}}}{e^{-\frac{s\pi}{2k}} - e^{-\frac{s\pi}{2k}}} \\ &= \frac{2\cosh\left(\frac{s\pi}{2k}\right)}{2\sinh\left(\frac{s\pi}{2k}\right)} \\ &= \coth\left(\frac{s\pi}{2k}\right) \end{aligned}$$

Finally,

$$\begin{aligned} L\left[\left|\sin kt\right|\right] &= \frac{k}{s^2+k^2}\left[\frac{1+e^{-\frac{s\pi}{k}}}{1-e^{-\frac{s\pi}{k}}}\right] \\ &= \frac{k}{s^2+k^2}\coth\left(\frac{s\pi}{2k}\right) \end{aligned}$$

Question 5.(10%) Consider Bessel's equation of order zero

$$x\frac{d^2y}{dx^2} + \frac{dy}{dx} + xy = 0$$

Using the method of series, it can be shown that one (continuously differentiable) solution of this equation is the *Bessel function of order zero*, denoted by $J_0(x)$. Find a second linearly independent solution (you may leave your solution in terms of an integral and the function $J_0(x)$).

Solution. We use the method of reduction of order (Section 3.4). The general solution of the differential equation is given by

$$y(x) = v(x) J_0(x)$$

where $v(x)$ is an unknown function to be determined.

$$\begin{aligned} \frac{dy}{dx} &= v' J_0 + v J_0' \\ \frac{d^2y}{dx^2} &= v'' J_0 + 2v' J_0' + v J_0'' \end{aligned}$$

Substituting into the differential equation, we obtain

$$\left(v'' J_0 + 2v' J_0'\right) + v' J_0 + v\left(xJ_0'' + J_0' + xJ_0\right) = 0$$

Since J_0 is known to solve the differential equation, we have that

$$xJ_0'' + J_0' + xJ_0 = 0$$

Thus, we obtain the reduced equation

$$xv'' J_0 + v'\left(2xJ_0' + J_0\right) = 0$$

Let $w = v'$.

$$w' + \left(2\frac{J_0}{J_0'} + \frac{1}{x}\right) w = 0$$

This equation is first order linear in w (see Section 2.5). An integrating factor is given by

$$\begin{aligned} \mu(x) &= \exp\left(2\ln|J_0| + \ln|x|\right) \\ &= \exp\left(\ln\left[|x| J_0^2\right]\right) \end{aligned}$$

Choose

$$\mu(x) = xJ_0^2$$

The differential equation in w now becomes

$$\begin{aligned} \frac{d}{dx}\left[wxJ_0^2\right] &= 0 \\ wxJ_0^2 &= c_1 \\ w(x) &= \frac{c_1}{xJ_0^2} \end{aligned}$$

Hence,

$$\begin{aligned} v(x) &= \int w(x)\,dx \\ &= \int \frac{c_1}{xJ_0^2(x)} dx + c_2 \end{aligned}$$

Finally, the general solution of the differential equation is given by

$$\begin{aligned} y(x) &= v(x) J_0(x) \\ &= c_1 J_0(x) \int \frac{dx}{xJ_0^2(x)} + c_2 J_0(x) \end{aligned}$$

so that a second linearly independent solution if given by

$$y_2(x) = J_0(x) \int \frac{dx}{xJ_0^2(x)}$$

Question 6.(15%) Find the general solution of the system

$$\begin{aligned}\frac{dx}{dt} &= 3x + 2y + z \\ \frac{dy}{dt} &= -x - z \\ \frac{dz}{dt} &= x + y + 2z\end{aligned}$$

Solution. Write the system in matrix form and use the theory of eigenvalues and eigenvectors in Section 3.10.

$$\begin{bmatrix} x' \\ y' \\ z' \end{bmatrix} = \begin{bmatrix} 3 & 2 & 1 \\ -1 & 0 & -1 \\ 1 & 1 & 2 \end{bmatrix} \begin{bmatrix} x \\ y \\ z \end{bmatrix}$$

The eigenvalues of the matrix

$$\begin{bmatrix} 3 & 2 & 1 \\ -1 & 0 & -1 \\ 1 & 1 & 2 \end{bmatrix}$$

are given by the determinant equation

$$\begin{aligned}\begin{vmatrix} 3-\lambda & 2 & 1 \\ -1 & -\lambda & -1 \\ 1 & 1 & 2-\lambda \end{vmatrix} &= 0 \\ -8\lambda + 5\lambda^2 + 4 - \lambda^3 &= 0 \\ (\lambda - 1)(\lambda - 2)^2 &= 0 \\ \lambda &= 2 \text{ (twice)}, 1\end{aligned}$$

Hence, the eigenvalues are given by

$$\lambda_1 = 1, \lambda_2 = 2 \text{ (twice)}$$

Hence, both eigenvalues are real, one is distinct and the other has multiplicity two. Next, find the corresponding eigenvectors.

$\lambda_1 = 1:$

Let the corresponding eigenvector be denoted by $\mathbf{v}_1 = \begin{bmatrix} c_1 & c_2 & c_3 \end{bmatrix}^T$ where the c_i, $i = 1, 2, 3$ are constant. Then, $\mathbf{v}_1$ is given by

$$\begin{bmatrix} 2 & 2 & 1 \\ -1 & -1 & -1 \\ 1 & 1 & 1 \end{bmatrix} \begin{bmatrix} c_1 \\ c_2 \\ c_3 \end{bmatrix} = \begin{bmatrix} 0 \\ 0 \\ 0 \end{bmatrix}$$

This is equivalent to the system

$$\begin{aligned} 2c_1 + 2c_2 + c_3 &= 0 \\ c_1 + c_2 + c_3 &= 0 \end{aligned}$$

which has solution

$$\begin{aligned} c_3 &= 0 \\ c_1 &= \text{arbitrary} \\ c_2 &= -c_1 \end{aligned}$$

Hence, the corresponding eigenvector is given by

$$\mathbf{v}_1 = \begin{bmatrix} 1 \\ -1 \\ 0 \end{bmatrix}$$

$\lambda_2 = 2:$

Since this eigenvalue is repeated, we will obtain either two linearly independent eigenvectors directly (as in Case 1 of Section 3.10) or only one eigenvector directly and a second linearly independent eigenvector from Equation (3.10.12) (as in Case 2 of Section 3.10). To see which of these cases applies here, let a corresponding eigenvector be denoted by $\mathbf{v}_2 = \begin{bmatrix} c_1 & c_2 & c_3 \end{bmatrix}^T$ where the c_i, $i = 1, 2, 3$ are constant. Then, $\mathbf{v}_2$ is given by

$$\begin{bmatrix} 1 & 2 & 1 \\ -1 & -2 & -1 \\ 1 & 1 & 0 \end{bmatrix} \begin{bmatrix} c_1 \\ c_2 \\ c_3 \end{bmatrix} = \begin{bmatrix} 0 \\ 0 \\ 0 \end{bmatrix}$$

This is equivalent to the system

$$\begin{aligned} c_1 + 2c_2 + c_3 &= 0 \\ c_1 + c_2 &= 0 \end{aligned}$$

which has solution

$$\begin{aligned} c_2 &= -c_1 \\ c_1 &= \text{arbitrary} \\ c_3 &= c_1 \end{aligned}$$

Hence, we have only one eigenvector, given by

$$\mathbf{v}_2 = \begin{bmatrix} 1 \\ -1 \\ 1 \end{bmatrix}$$

To find a second linearly independent eigenvector $\mathbf{v}_3$ associated with $\lambda_2 = 2$, we follow the method outlined in Case 2 of Section 3.10, that is, we solve Equation (3.10.12)

$$(A - \lambda_2 I)\,\mathbf{v}_3 = \mathbf{v}_2$$

with $\lambda_2 = 2$:

$$\begin{bmatrix} 1 & 2 & 1 \\ -1 & -2 & -1 \\ 1 & 1 & 0 \end{bmatrix} \mathbf{v}_3 = \begin{bmatrix} 1 \\ -1 \\ 1 \end{bmatrix}$$

Let the corresponding eigenvector be denoted by $\mathbf{v}_3 = \begin{bmatrix} c_1 & c_2 & c_3 \end{bmatrix}^T$ where the c_i, $i = 1, 2, 3$ are constant. Then, $\mathbf{v}_3$ is given by

$$\begin{bmatrix} 1 & 2 & 1 \\ -1 & -2 & -1 \\ 1 & 1 & 0 \end{bmatrix} \begin{bmatrix} c_1 \\ c_2 \\ c_3 \end{bmatrix} = \begin{bmatrix} 1 \\ -1 \\ 1 \end{bmatrix}$$

This is equivalent to the system

$$\begin{aligned} c_1 + 2c_2 + c_3 &= 1 \\ c_1 + c_2 &= 1 \end{aligned}$$

which has solution

$$\begin{aligned} c_1 &= 1 - c_2 \\ c_3 &= 1 - c_1 - 2c_2 \\ &= 1 - (1 - c_2) - 2c_2 \\ &= -c_2 \\ c_2 &= \text{arbitrary} \end{aligned}$$

Hence, the corresponding eigenvector (choosing $c_2 = 1$) is given by

$$\mathbf{v}_3 = \begin{bmatrix} 0 \\ 1 \\ -1 \end{bmatrix}$$

The general solution of the system is now given by (see (3.10.7))

$$\begin{aligned} \mathbf{x}(t) &= \begin{bmatrix} x \\ y \\ z \end{bmatrix} \\ &= c_1 \mathbf{v}_1 e^{\lambda_1 t} + c_2 \mathbf{v}_2 e^{\lambda_2 t} + c_3 e^{\lambda_2 t}\left(\mathbf{v}_2 t + \mathbf{v}_3\right) \\ &= c_1 \begin{bmatrix} 1 \\ -1 \\ 0 \end{bmatrix} e^{t} + c_2 \begin{bmatrix} 1 \\ -1 \\ 1 \end{bmatrix} e^{2t} + c_3 \left(\begin{bmatrix} 1 \\ -1 \\ 1 \end{bmatrix} t + \begin{bmatrix} 0 \\ 1 \\ -1 \end{bmatrix} \right) e^{2t} \end{aligned}$$

or

$$\begin{aligned} x(t) &= c_1 e^t + c_2 e^{2t} + c_3 t e^{2t} \\ y(t) &= -c_1 e^t - c_2 e^{2t} + c_3 (1-t) e^{2t} \\ z(t) &= c_2 e^{2t} + c_3 (t-1) e^{2t} \end{aligned}$$

Question 7.(10%) Solve the following initial value problem.

$$\begin{aligned} e^{2x} \frac{d^2 x}{dt^2} &= -1, \qquad t > 0, \ \frac{dx}{dt} > 0 \\ x(0) &= 0, \ \frac{dx}{dt}(0) = 1 \end{aligned}$$

Solution. The differential equation is one of the class in which the independent variable t does not appear explicitly (see Example 3.7.2). Let

$$\begin{aligned} p &= \frac{dx}{dt} \\ \frac{d^2 x}{dt^2} &= \frac{dp}{dt} = \frac{dp}{dx}\frac{dx}{dt} = p\frac{dp}{dx} \end{aligned}$$

The differential equation becomes

$$e^{2x} p \frac{dp}{dx} = -1$$

which is *separable* (see Section 2.1). Separating the variables, we obtain

$$\begin{aligned} \int p dp &= -\int e^{-2x} dx \\ \frac{p^2}{2} &= -\left(\frac{e^{-2x}}{-2}\right) + c_1 \\ &= \frac{1}{2e^{2x}} + c_1 \end{aligned}$$

Applying the initial conditions that $\frac{dx}{dt}(0) = p(0) = 1$ and $x(0) = 0$, we obtain

$$\begin{aligned} \frac{1}{2} &= \frac{1}{2} + c_1 \\ c_1 &= 0 \end{aligned}$$

Hence,

$$\begin{aligned} p^2 &= e^{-2x} \\ p &= \frac{dx}{dt} = e^{-x} \ (\text{since } p = \frac{dx}{dt} > 0) \end{aligned}$$

This differential equation is again separable.

$$\begin{aligned} \int e^x dx &= \int dt \\ e^x &= t + c_2 \end{aligned}$$

Again, since $x(0) = 0$, $c_2 = 1$. Finally, the solution is given by

$$\begin{aligned} e^x &= (t+1) \\ x(t) &= \ln(t+1), \; t > 0 \end{aligned}$$

APPENDIX

Differential Calculus

In what follows, f, g and h are sufficiently smooth functions of x. The derivatives of some of the more common functions arising in calculus are listed in the following table.

Table A.1 Common Derivatives

$$\frac{d}{dx}(C) = 0$$
$$\frac{d}{dx}(x^n) = nx^{n-1}$$
$$\frac{d}{dx}(\sin x) = \cos x$$
$$\frac{d}{dx}(\cos x) = -\sin x$$
$$\frac{d}{dx}(\tan x) = \sec^2 x$$
$$\frac{d}{dx}(\cot x) = -\csc^2 x$$
$$\frac{d}{dx}(\sec x) = \sec x \tan x$$
$$\frac{d}{dx}(\csc x) = -\csc x \cot x$$
$$\frac{d}{dx}(\ln x) = \frac{1}{x}, \; x > 0$$
$$\frac{d}{dx}(e^x) = e^x$$
$$\frac{d}{dx}(\arcsin x) = \frac{1}{\sqrt{1-x^2}}$$
$$\frac{d}{dx}(\arccos x) = -\frac{1}{\sqrt{1-x^2}}$$
$$\frac{d}{dx}(\arctan x) = \frac{1}{1+x^2}$$
$$\frac{d}{dx}(\operatorname{arccot} x) = -\frac{1}{1+x^2}$$
$$\frac{d}{dx}(\operatorname{arcsec} x) = \frac{1}{x\sqrt{1-x^2}}$$
$$\frac{d}{dx}(\operatorname{arccsc} x) = -\frac{1}{x\sqrt{1-x^2}}$$

Here, C and n are constants and all arguments of the trigonometric functions are in radians.

Product Rule

$$\frac{d}{dx}(f(x)g(x)) = f(x)\frac{d}{dx}(g(x)) + g(x)\frac{d}{dx}(f(x))$$

Quotient Rule

$$\frac{d}{dx}\left(\frac{f(x)}{g(x)}\right) = \frac{g(x)\frac{d}{dx}(f(x)) - f(x)\frac{d}{dx}(g(x))}{(g(x))^2}$$

Integral Calculus

Integration by Parts

This technique is used to integrate products or quotients or functions.

For *indefinite* integrals:

$$\int f(x)g'(x)dx = f(x)g(x) - \int g(x)f'(x)dx + C$$

Alternatively, replace $g'(x)$ with some function $h(x)$, (and hence $g(x)$ by *any* antiderivative of $h(x)$, say $H(x)$), to obtain

$$\int f(x)h(x)dx = f(x)H(x) - \int H(x)f'(x)dx + C$$

For *definite* integrals:

$$\int_a^b f(x)h(x)dx = [f(x)H(x)]_a^b - \int_a^b H(x)f'(x)dx$$

Integration by Substitution

The substitution

$$\begin{aligned} u &= g(x) \\ du &= g'(x)\,dx \end{aligned}$$

is used to simplify the integral under consideration. The most difficult (and important) part of this procedure lies in the correct choice of the function $g(x)$.

There is no general rule for choosing the correct substitution. However, there are certain cases when the particular form of the integral in question suggests the best choice of $g(x)$. For example, consider an integral of the form

$$\int f(g(x))g'(x)dx \tag{A1}$$

Let $u = g(x)$. Since

$$\frac{du}{dx} = g'(x)$$

the differential dx is related to the differential du by

$$du = g'(x)dx$$

or

$$dx = \frac{du}{g'(x)}$$

The integral now becomes

$$\int f(u)du$$

which (hopefully), is simpler to integrate than the original integral (A1). Once the (simplified) integral is determined (in terms of the variable u) - from tables or otherwise, the substitution $u = g(x)$ is used to express the answer in terms of the original variable x. Hence, if an integral is of the form (A1), the most appropriate substitution is $u = g(x)$.

To identify an integral as being of the form (A1), look for a function $g(x)$ whose differential $g'(x)dx$ appears in the integral. The function $g(x)$ need not appear alone, it may be part of a composite function $f(g(x))$ as in (A1). However, the differential $g'(x)dx$ must appear alone and not as part of another function. Once $g(x)$ is identified, the correct substitution is $u = g(x)$. For example, for the integral

$$\int \frac{\cos x}{1 + \sin^2 x} dx$$

is of the form (A1) with $u = g(x) = \sin x$ and $f(u) = \dfrac{1}{1 + u^2}$.

As a last resort, choose the substitution $u = g(x)$ where $g(x)$ is some complicated part of the integral. Again, there are no guarantees, but in many cases, the integral does simplify considerably.

Table A.2 Table of Integrals

$$\int k dx = kx + C$$

$$\int x^n dx = \frac{x^{n+1}}{n+1} + C, \quad n \neq -1$$

$$\int \frac{1}{x} dx = \ln|x| + C$$

$$\int (ax+b)^n dx = \begin{cases} \frac{1}{a}\left(\frac{(ax+b)^{n+1}}{n+1}\right) + C, & n \neq -1 \\ \frac{1}{a}\ln|ax+b| + C, & n = -1 \end{cases}$$

$$\int e^{kx} dx = \frac{e^{kx}}{k} + C, \quad k \neq 0$$

$$\int x e^{kx} dx = \frac{e^{kx}(kx-1)}{k^2} + C, \quad k \neq 0$$

$$\int e^{ax} \sin bx dx = \frac{e^{ax}}{a^2+b^2}(a \sin bx - b \cos bx) + C$$

$$\int e^{ax} \cos bx dx = \frac{e^{ax}}{a^2+b^2}(a \cos bx + b \sin bx) + C$$

$$\int k^{ax} dx = \frac{k^{ax}}{a \ln k} + C, \quad k > 0,\ a \neq 0$$

$$\int \ln x dx = x \ln x - x + C, \ x > 0$$

$$\int x^m \ln x dx = \frac{x^{m+1}}{(m+1)^2}[(m+1)\ln x - 1] + C, \ x > 0$$

$$\int \sin x dx = -\cos x + C$$

$$\int \sin(ax+b) dx = -\frac{1}{a}\cos(ax+b) + C$$

$$\int \cos x dx = \sin x + C$$

$$\int \cos(ax+b) dx = \frac{1}{a}\sin(ax+b) + C$$

$$\int \tan x dx = \ln|\sec x| + C$$

$$\int \cot x dx = \ln|\sin x| + C$$

$$\int \sec x dx = \ln|\sec x + \tan x| + C$$

Table A.2 Continued

$$\begin{aligned}
\int \csc x dx &= \ln|\csc x - \cot x| + C \\
\int \frac{dx}{x^2+k^2} &= \frac{1}{k}\arctan(\frac{x}{k}) + C, \ k > 0 \\
\int \frac{dx}{\sqrt{k^2-x^2}} &= \arcsin(\frac{x}{k}) + C, \quad k \neq 0 \\
\int \sin^2 x dx &= \frac{x}{2} - \frac{\sin 2x}{4} + C \\
\int \cos^2 x dx &= \frac{x}{2} + \frac{\sin 2x}{4} + C \\
\int \tan^2 x dx &= \tan x - x + C \\
\int kf(x)dx &= k\int f(x)dx \\
\int [f(x)+g(x)]dx &= \int f(x)dx + \int g(x)dx \\
\int \frac{h'(x)}{h(x)}dx &= \ln|h(x)| + C
\end{aligned}$$

Here, k, a, b and n are real constants, m is a positive integer and C is an arbitrary constant (of integration).

Partial Fractions

Case 1

If $\frac{P(x)}{Q(x)}$ is a *proper rational function*, that is, P and Q are polynomials such that the *degree of* P *is less than that of* Q, then $\frac{P(x)}{Q(x)}$ decomposes into partial fractions as follows.

$$\begin{aligned}
\frac{P(x)}{Q(x)} &= \frac{P(x)}{(a_1x+b_1)(a_2x+b_2)+\cdots+(a_kx+b_k)} \\
&= \frac{A_1}{a_1x+b_1} + \frac{A_2}{a_2x+b_2} + \cdots + \frac{A_k}{a_kx+b_k}
\end{aligned}$$

The constants $A_1, ..., A_k$ are to be determined.

Case 2

If one of the factors, for example, (a_1x+b_1), in the denominator Q of the proper rational function $\frac{P}{Q}$, is repeated, say, r times, (i.e. the factor $(a_1x+b_1)^r$ appears in the factored form of $Q(x)$) then, instead of the single partial fraction $\frac{A_1}{a_1x+b_1}$, assign , r partial fractions as follows.

$$\frac{P(x)}{(a_1x+b_1)^r}=\frac{A_1}{a_1x+b_1}+\frac{A_2}{(a_2x+b_2)^2}+\cdots+\frac{A_r}{(a_rx+b_r)^r}$$

The constants $A_1,...,A_r$ are to be determined.

Case 3

If, in the expression $\frac{P}{Q}$, Q contains an *irreducible quadratic* factor (i.e. a quadratic factor which cannot be factored into 2 real linear factors - more precisely, a factor of the form ax^2+bx+c where $b^2-4ac<0$) the corresponding partial fraction will take the form

$$\frac{Ax+B}{ax^2+bx+c}$$

If that factor is repeated, say, r times, then assign r partial fractions as follows.

$$\frac{P(x)}{(ax^2+bx+c)^r}$$
$$=\frac{A_1x+B_1}{ax^2+bx+c}+\frac{A_2x+B_2}{(ax^2+bx+c)^2}+\cdots+\frac{A_rx+B_r}{(ax^2+bx+c)^r}$$

The constants $A_1,...,A_r$, $B_1,...,B_r$ are to be determined.

Exponential and Logarithmic Functions

Properties of the Natural Exponential Function $f(x)=e^x$

Let x and y be real numbers.

$$\begin{aligned}
y &= f(x)=e^x>0\\
e^0 &= 1\\
e^{-x} &= \frac{1}{e^x}\\
e^{(x+y)} &= (e^x)(e^y)\\
e^{(x-y)} &= (e^x)(e^{-y})=\frac{e^x}{e^y}
\end{aligned}$$

Properties of the Natural Logarithmic Function $f(x) = \ln x, \; x > 0$

The natural logarithmic function, $f(x) = \ln x$, is defined as the *inverse* of the natural exponential function. That is,

$$x = e^y \iff y = \ln x$$

Since $x = e^y > 0$ (for any real number y), it follows that the natural logarithmic function is defined *only for real numbers* $x > 0$. That is,

$$y = f(x) = \ln x, \qquad x > 0$$

Let x, y and r be real numbers.

$$\begin{aligned}
\ln e^x &= x \\
e^{\ln x} &= x, \quad x > 0 \\
\ln 1 &= 0 \\
\ln e &= 1 \\
\ln(xy) &= \ln x + \ln y, \qquad x, y > 0 \\
\ln(\frac{x}{y}) &= \ln x - \ln y, \qquad x, y > 0 \\
\ln x^r &= r \ln x, \qquad x > 0 \\
-\infty &< \ln x < \infty
\end{aligned}$$

Completing the Square

Let b and c be real numbers.

$$x^2 + bx + c = \left(x + \frac{b}{2}\right)^2 - \frac{b^2}{4} + c$$

Other titles in the *Smart Practices* series:

Calculus Solutions: How to Succeed in Calculus
0 13 287475 X

Ordinary Differential Equations
0 13 907338 8

Coming in 1998:

Statistics

Linear Algebra